:(

Other Pergamon publications of related interest

PARSONS, TAKAHASHI & HARGRAVE
  Biological Oceanographic Processes, 3rd Edition

PURCHON
  The Biology of the Mollusca, 2nd Edition

RAYMONT
  Plankton and Productivity in the Oceans, 2nd Edition
    Volume 1 - Phytoplankton
    Volume 2 - Zooplankton

FISH AQUACULTURE
Technology and Experiments

by

CHRISTOPH MESKE

*Federal Research Centre for Fisheries,
Institute for Coastal and Inland Fisheries,
Hamburg/Ahrensburg, Federal Republic of Germany*

Edited and translated by

FREDERICK VOGT

*Formerly of the Polytechnic of Central London,
London, UK*

PERGAMON PRESS
OXFORD · NEW YORK · TORONTO · SYDNEY · PARIS · FRANKFURT

U.K.	Pergamon Press Ltd., Headington Hill Hall, Oxford OX3 0BW, England
U.S.A.	Pergamon Press Inc., Maxwell House, Fairview Park, Elmsford, New York 10523, U.S.A.
CANADA	Pergamon Press Canada Ltd., Suite 104, 150 Consumers Road, Willowdale, Ontario M2J 1P9, Canada
AUSTRALIA	Pergamon Press (Aust.) Pty. Ltd., P.O. Box 544, Potts Point, N.S.W. 2011, Australia
FRANCE	Pergamon Press SARL, 24 rue des Ecoles, 75240 Paris, Cedex 05, France
FEDERAL REPUBLIC OF GERMANY	Pergamon Press GmbH, Hammerweg 6, D-6242 Kronberg-Taunus, Federal Republic of Germany

First edition 1985

British Library Cataloguing in Publication Data
Meske, Christoph
Fish Aquaculture.
1. Fish—Culture
I. Title II. Vogt, F.
III. Aquakultur. *English*
639.3   SH151
ISBN 0 08 024920 5 (Hardcover)
ISBN 0 08 024919 1 (Flexicover)

Printed in Great Britain by A. Wheaton & Co. Ltd., Exeter

EDITOR'S AND TRANSLATOR'S NOTE

As an ex-farm manager I got into fish farming by accident while researching the use of renewable energies in the bio-industries as a member of Prof Ray Maw's Built Environment Research Group at the Polytechnic of Central London.

In the process I made a lot of friends: (Dr) Ann M Powell of the City Polytechnic; (Dr) Alan Starkie of the Severn and Trent Water Authority (who first brought Prof Meske's book to my attention) and (Dr) Colin Purdom of the Fisheries Laboratory (Ministry of Agriculture, Fisheries & Food). Unfortunately I met (Prof) John Bardach of the East West Centre and the University of Hawai too late to avail myself of his kind offer of help.

Without their willing help, advice and constant encouragement I would not have considered tackling Prof Meske's work nor would I have felt able to do so. I owe them a great debt.

Without Prof Maw's early encouragement this translation would never have gone beyond the initial stages.

Prof Meske's original book has been updated and augmented and it is fitting here to thank Pergamon Press, particularly Jill Price, Martin Richardson and John Cooper, for their support in getting it published.

As an agriculturist I found fish farming a primitive industry which could and should make a major contribution to the provision of protein for food and job creation for society. These aims can realistically only be achieved by fundamental changes in the structure of the industry along the lines of other livestock enterprises and certain arable crops where stock selection, management, marketing and financing developments have given large gains and stability to both producers and consumers.

F. Vogt

PREFACE

Aquaculture has gained a momentum throughout the world during recent decades which is probably unparalleled in other branches of food production. Cropping fish and other aquatic organisms for food is to a large extent based on very old methods. All the traditional methods together with their modern counterparts are often of high technology and are now commonly described as aquaculture.

In most countries of the world there is a surge of activity in this field. There are new research centres, journals, textbooks and many international meetings. All of these confirm the topical interest in the subject of aquaculture. In recent years it has become clear that the problems of overpopulation, of food and raw material procurement and of environmental pollution demand major research work.

This book is an enlarged and updated version of my original work which was first published by Verlag Ulmer of Stuttgart in 1973.

It describes in Chapters 2 and 3 the methods currently used for the production of those warm water table fish which are of major importance. The main part of the book — Chapters 4, 5 and 6 — describes the experiments and procedures which may help to combat the growing food problem through new production methods for animal protein. The aim of the work presented here concerns the continuous production of warm water table fish independently of climate or environment within the least necessary space and even in regions with unsuitable weather or topography.

A large number of experiments were carried out to gain basic knowledge on the physiological reactions of fish to various environmental influences. In particular the investigations were concerned with the effect of abiotic factors and nutrition on the growth of carp. It is hoped that this book may be of practical use to commercial fish farmers. Perhaps it will also make a small contribution towards solving some of the problems of food production and environmental pollution.

Although the development of a closed cycle or re-circulation system was primarily designed for experimental work, there was an underlying thought of its possible use in large commercial enterprises.

All the work described in this book was undertaken at Ahrensburg, near Hamburg. During the early years up to 1970 it took place at the Max Planck Institute for Plant Breeding and from 1971 onwards at the Institute for Coastal and Inland Fisheries of the Federal German Fisheries Research Institute.

Although the work reported in this book refers solely to a closed cycle system using the activated sludge process, this system represents only one of several alternatives. Many centres throughout the world are working on the development of re-circulation systems and it is inevitable that this will lead to improved and more economic system performance. As the availability of clean water tends to decrease, if only regionally, it is difficult to see how fish farming can be carried on without re-circulation. Fish farming is a growing industry. It demands an uninterrupted supply of clean water. It would seem that re-circulation systems may be the only answer.

The work could not have been carried out without my co-workers' unstinting support and advice with the experiments and the construction and maintenance of the plant.

I thank them all most sincerely for their help, particularly when willingly carried out over week-ends and throughout the holidays. Without them the research work — and the publication of this book — would not have been possible. The photographs have been prepared by Mr K. Engelhardt to whom I express my special thanks.

Finally it is appropriate that I thank the editor and translator for his endeavours.

It is hoped that this book will make a small contribution towards easing the problems of food production and of pollution.

Ahrensburg 1982   Christoph Meske

CONTENTS

1. INTRODUCTION

Fish for human consumption — whether eaten directly, processed or fed to domestic animals as fish meal — is almost entirely harvested from the seas. In 1980 marine catches amounted to 64.7 million tons (Mt). Inland fishing provided an additional 7.6 Mt of which roughly half was produced by both salt and fresh water aquaculture. This figure can be broken down into 3.23 Mt of fish, 3.7 Mt of molluscs and 0.71 Mt of crustaceans.

The fishing industry has undergone considerable changes in recent times, relying increasingly on higher technology, improved methods and more effective gear. Nevertheless it has remained basically the same: hunting for wild fish in natural waters.

During recent years the problems of the fishing industry have grown considerably. Overfishing has resulted in restrictions and quotas. Almost all coastal states now operate 200 mile (320 km) limits. There are total bans on the fishing of some species, while other previously little or unknown species have appeared on the market. The bans have affected those countries traditionally engaged in high seas fishing. As a result joint ventures between the technically advanced nations and the underdeveloped countries have been inaugurated in the fields of research, processing and marketing.

Apart from overfishing there is also the problem of pollution of which oil slicks and toxic industrial waste effluent are examples. In recent oil rig disasters large quantities of oil were discharged into the sea. There have been several major tanker accidents. In Japan industrial discharge caused a new medical condition.

A number of other factors influence the fish crop adversely. Cannibalism, parasites, viral and bacterial infection occur naturally. As the natural balance is upset by man-made changes in the environment the consequential and cumulative effect may have serious repercussions on the world's food supply.

OVERFISHING

The history of post-war fishing is one of increasing catches. In the 11 years from 1957 to 1968 catches roughly doubled and rose from 40 Mt a year in 1960 to 63.3 Mt a year in 1968. The first sign of a turndown appeared in 1969 when there was a 1.2 Mt or 2% reduction in the catch on the previous year.

Up to this time it was thought that the global crop would continue to increase at the same rate as before. Catches rose as fishing fleets equipped themselves with more advanced technical aids and gear, while at the same time reducing the chances of the fish population to regenerate itself. Between 1969 and 1974 the fishing fleet grew by over 50%, consisting of larger vessels and factory ships. It is open to debate whether the arrest in the crop yield during this period is due to the expansion of the fleet, or whether even the greater number of ships were unable to raise the level of catches (Table 1).

TABLE 1.

World Catches 1956–1980 (FAO) in million metric tonnes

1956	30.5	1968	64.3
1958	33.2	1970	69.5
1960	40.0	1972	65.5
1962	46.9	1974	70.3
1964	52.8	1976	69.6
1966	57.4	1978	70.4
		1980	72.3

The FAO made an optimistic forecast of improving catches rising to 130 Mt a year in the year 2000. This figure is open to doubt, particularly as the increasing sophistication of the equipment is likely to expose the residual population in the coastal water of the developing countries to the dangers of overfishing. Table 1 shows that world fish catches have not risen to any extent during the last 10 years.

When Iceland first imposed its 50 mile (80 km) fishing limit there was considerable diplomatic activity followed by the cod war with the United Kingdom. The International Court at The Hague was unable to exert its authority. Since then 200 mile (320 km) fishing limits have become the norm and the smaller seas, like the Baltic, have been divided among bordering states. The stress caused within the European Economic Community (EEC) over national or traditional fishing rights gives an indication of the political importance now given to fishing. The freedom of the seas is being less frequently mentioned.

Fishing has until now been based exclusively on economic criteria while neglecting biology and the environment. The very profitable but biologically lethal exploitation of the cod spawning shoals is an example, as is the catching of immature herring for feeding to trout. It should be noted that it takes 6 kg of wet herring to produce 1 kg of trout. As a result of over-fishing there was for a period a total ban on catching herring in the North Sea except for fish for direct human consumption in Danish coastal waters (Table 2). While total world catches

TABLE 2.

North Sea herring Catches in million tons

1965	1.2	1977	0.084
1975	0.313	1978	0.030
1976	0.190	1980	nil

remained steady there was a rapid decline in the anchovy crop from the west coast of South America. Anchovy catches off Peru dropped from its place as the leading fishmeal producer with 2.25 Mt in 1970 to fourth place with only 0.44 Mt in 1977. The disappearance of the anchovy is attributed partly to a change in the Humboldt current and partly to overfishing.* Peru consequently introduced conservation measures. Globally fish meal production fell by 9.5% between 1976 and 1977. Fish meal remains the most important component in compound feeds for pigs, poultry and table fish.

While the dangers of overfishing have been generally accepted, conservation measures have been slow in being implemented. There are now international agreements on closed seasons, regional quotas and total bans covering some of the important species of fish. Controlled production and rearing of marine fish for seeding the seas may lead to a slow rise in the yields, provided that the research work now being carried out at many centres is stepped up. Through controlled management, feeding and rearing of marine fish a bridge between fishing and aquaculture can be established. Work in this relatively new research field appears to be most advanced in Japan where there is a plan to convert large areas of inland seas into the equivalent of meadows in which stock can graze. The basis of this plan rests on the considerable successes Japanese workers have had with the controlled production of salt water fry such as red sea bream (*Chrysophrys major*), shrimp and mollusc larvae. The project also includes pathology and genetics.

Japan has the advantage of a coast with many bays. This allows large areas of sea to be exploited by planned developments. Similar schemes in less-favoured locations are likely to become important in the future. The successful work on flat fish carried out in Great Britain and the culture of salmon in Norway are indications of what can be achieved.

POLLUTION

Both marine and fresh water fish are increasingly exposed to toxic matter arising directly or indirectly from man-made industrial waste, both sea- and land-borne. This includes industrial

---

* Editor's note: The recurring Pacific El Nino weather system alters the weather and currents of the region. The El Nino storms of 1972 are thought to have wiped out nearly 90% of the anchovy fishing industry along the South American coast. The anchovy population might have recovered but for the excessive fishing. In 1982/83 the system is assumed to have been responsible for the population crash affecting 17 million birds — terns, shearwaters, frigates and other species. The birds abandoned their nesting sites when their food sources failed. The El Nino changes were first described by Sir Gilbert Walker of the Indian Weather Service some 50 years ago. He called them southern oscillations.

effluent, sewage discharged into rivers, lakes and the sea as well as bilge. The discharge may be continuous or sporadic and the effect of the pollution on commercially exploited fish can be divided into three groups:

1. *Sudden wipe-out of the fish from acute poisoning due to the addition of large amounts of highly toxic substances.*

As an example the River Rhine disaster of 1969 can be cited. Fish deaths were catastrophic. Initially it was presumed that an insecticide, toxaphene, was the cause, but its connection with the disaster remained unproven. Many similar cases are being reported without the cause being determined in the majority of cases. The drainage or leaching from farmed land is known to be responsible for cases of wipe-outs in rivers and lakes. American work indicates that mixtures of certain chemicals are often considerably more lethal than component substances alone (Marking and Dawson, 1975).

2. *The effect of toxic substances on eggs and larvae.*

This is of special importance in coastal waters in which commercially important fish tend to develop but where they are exposed to fluvial detritus. During the early stages of life fish — like all higher organisms — are particularly susceptible to toxicity (Goetting, 1971). Destruction at this stage has a rapid effect on the size of the fish population although when adult these fish may be tolerant to a certain low level of toxicity.

3. *Storage of toxic substances at sub-lethal levels in commercial fish.*

Best known examples are the accumulation of mercury and insecticides such as DDT. Again, arising from industrial and agricultural pollution these substances are carried in the rivers to the sea where they enter the food chain via plankton. Eventually they end up in carnivores, such as the tunny, swordfish and pike, as well as in birds of prey, such as the sea eagle, and mammals, e.g. seals. In the case of organomercuric compounds high concentration is rapidly achieved because carnivores find it easy to catch affected fish whose responses are impeded by the poison. These carnivores often have high contamination levels as a consequence. A sick seal caught in Lake Saimaa in Finland had a muscular mercury content of 197 ppm (Nuorteva, 1971). In southern California seals were found to have up to 700 ppm of mercury in their livers (Anas, 1974). Among marine fish the swordfish (*Xiphias gladius*) and the marlin (*Makaira indica*) appear to be the species most contaminated by mercury with an average of 7.3 ppm intramuscular mercury (MacKay *et al.*, 1975). The permissible limit in the USA and Canada is 0.5 ppm.

The first recorded case of human disease and deaths caused by consumption of mercury-contaminated fish occurred in Japan: Minamata Disease. The fish eaten contained 10 ppm of mercury and drew attention to heavy metal contamination (e.g. Pearson and Frangipane, 1975). In Europe too, especially in Scandinavia, high mercury values were observed in fish. Although no lethal cases were reported high concentrations of this metal were found in the blood and hair of people eating a lot of fish. Pike caught in several Finnish rivers contained mercury far above the limits of tolerance.

Apart from the heavy metals — mercury, lead, cadmium, chrome, zinc, nickel and copper — the following toxic residues in fish are of particular importance: non-metallic trace elements, oils and their derivatives, pesticides and herbicides.

The connection between industrial pollution and poisonous effluent in rivers and the contamination of fish has repeatedly been recorded (e.g. Reichenback-Klinke, 1976). There is the danger to the consumer that within the food chain they are contaminated by apparently healthy fish. According to Reichenbach-Klinke (1971) the concentration of poisons in the fish compared to that in the water is of the order of 1110 for aldrin and dieldrin, up to 400 for toxaphene and 8500 for DDT.

2. AQUACULTURE — THE CONCEPT

As agriculture is the cultivation of animals and plants and the cropping from the land so aquaculture is the pursuit of production from water. Aquaculture includes the husbandry, management, nutrution and multiplication or breeding of all useful aquatic organisms. Just as farming replaced hunting and gathering so aquaculture may increasingly displace the hunting and catching of wild stock. Aquaculture is mainly concerned with fish, crustaceans, molluscs and algae.

Some production methods are based on old if not ancient practices, especially in Asia. Others are the result of modern scientific research and experiments exploiting industrial techniques and new raw material resources.

Eighty per cent of all fish produced by aquaculture come from Asia. There the common carp and Chinese carp species are kept in small ponds, rice paddies, ditches or channels. They contribute significantly to the local food. In South-East Asia fish and molluscs provide over 50% of all consumed animal protein. China produces over 60% of the world's farmed fish. In India brackish water aquaculture developed as the result of coastal land reclamation. Today aquaculture's contribution to Indian fish production stands at 38% of the total — equivalent to 13% of global output. In North America the development of aquaculture is fairly recent and is concerned in the main with the production of high-value table fish and the stocking of fish for anglers.

In Europe the monks launched aquaculture during the Middle Ages, mostly by keeping carp in ponds. In recent times trout and other fish of commercial value are also being cultivated.

The doubling of the world's population by the year 2000 is forecast. Using improved techniques and management it should be possible to raise output of protein and carbohydrates through faster-growing, disease-resistant cereals, crop plants and livestock. As the population grows the pressure on available land increases and the search for alternative methods of food production will intensify. Biologically there is a limit to the size of the natural stock level of fish, molluscs and crustaceans. Total world catches currently hover around the 70 million tonnes a year mark and there is obviously a limit to which they can be further raised.

In contrast to the fishing industry there is a general consensus that aquaculture production is capable of considerable improvement. The FAO estimates that aquaculture output can be increased many times by the year 2000 through optimal use of existing resources and the application of science. It will be necessary to use presently unproductive areas such as swamps, marshes and uncultivatable or uneconomic land. Some of this land could be utilized by pond complexes. In many tropical countries there are large mangrove swamps which could be put to use in mariculture and brackish water enterprises.

MARICULTURE

The culture of marine organisms is confined to operations in coastal waters. Mariculture is concerned with multicellular algae, molluscs, crustaceans and fish and can be grouped under 5 headings:

1. SUBSTRATE SYSTEMS

In this method aquatic organisms are encouraged to grow on substrates. Vertical beds or ropes are suspended in the water from anchored floats or fixed structures to aid the growth of the young organism. The substrate sites are mostly located in bays or the mouths of rivers. The same principle is used for seaweed production, e.g. in Japan, and for the cultivation of molluscs (mussels and oysters) in many parts of the world.

2. SEAWATER PONDS

By means of simple sluices the tides may be harnessed to renew the water in ponds. In Asia the use of salt water ponds for producing fish and crustaceans is widespread. One outstanding example is shrimp farming in Japan.

3. CAGES

Under this heading fall all forms of mariculture in which stock is kept free-swimming within flexible or fixed-cage type systems. Marine fish kept in net cages are afforded a constant change of water. The cages contain salt water fish for rearing and feeding but are not used for breeding.

Up till now the fry of many fish species needed to be caught at sea. However it is now possible to breed and rear a number of species in tanks. For instance the red sea bream (*Chrysophrys major*) is spawned and reared in Japan. Net cages are used to a considerable extent in the Far East, particularly in the many bays of Japan. Here the yellowtail (*Seriola quinqueradiata*) is an example. In Scotland net cage husbandry of the salmon (*Salmo salar*) has been developed after the problem of algae growth on the nets had been solved and an optimal feeding regime researched.

4. ENCLOSURES

This method is used exclusively for fish along coastlines lending themselves to being fenced off or enclosed by netting. An example is the farming of salmon (*Salmo salar*) in sections of Norwegian fjords.

5. TANKS

Here pumped sea water circulates through tanks. The water may be warm after having passed through the cooling towers of a power station, a method used for instance in England and Scotland. Tanks allow excellent control of the management, feeding and breeding of the fish. The system is capital-intensive but most suitable for maricultural research. A lot depends on the quality of the water circulating through the tanks and it is therefore essential that a reliable system is developed. The description of an experimental circuit is given in Chapter 4.

AQUACULTURE IN FRESH AND BRACKISH WATER

This aspect of aquaculture can be divided into two groups:

1. Natural waters such as inland seas, lakes, rivers, lagoons and mangrove swamps.
2. Artificial or man-made facilities such as ponds.

The exploitation of inland seas and lakes is predominantly by fishing boats. The lakes of South America and Africa are examples. Here the indigenous fish species are caught by traditional methods. Planned fish production in lakes is possible by stocking and by cage husbandry. The stocking of traditionally fished waters can be considered to be the first stage of aquaculture.

A few years ago fish pens were developed in the Philippines (e.g. Laguna de Bay). Here fish, particularly the milk fish (*Chanos chanos*), are farmed in fenced-off pens without supplementary feeding.

Rivers of the tropics and near-tropics are often still fished by primitive methods. On smaller rivers, e.g. in Indonesia, fish farming in net cages can be found. Many rivers of South America and Africa are free from contamination and have fairly high constant temperatures. Making use of these natural assets in flow-through or by-pass ponds together with a balanced feeding programme offers advantages. Whether such a system is feasible in rivers with high tidal differentials, such as the Amazon, has yet to be established.

Lagoons are usually formed by sandbanks, embankments or similar structures. They are usually shallow and while partially separated from rivers or the sea, they are normally connected with them. The water is therefore predominantly brackish. In Africa and Asia lagoons are exploited for aquaculture. The massive areas of mangrove swamps, as for instance in West Africa, today make only a minor contribution to aquaculture. According to FAO estimates lagoons and mangrove swamps are of considerable potential to aquaculture. An area of between 10 and 20 million hectares, of which the Niger delta alone accounts for 400 000 ha, should be available for fish farming.

Ponds used for fish production are found throughout the world. The most intensive cropping takes place in North America (catfish, *Ictalurus* spp.), Europe (trout) and Asia (carp, *C. carpio*) and herbivorous species in China. The pond management in large parts of South America and Africa is comparatively undeveloped. The management of ponds is divided into three groups:

1. extensive;
2. semi-intensive;
3. intensive.

In extensive ponds fish are kept entirely without supplementary feed, relying completely on the pond fauna and flora. The productivity of these ponds can be improved by the application of fertilizer or manure. Extensive carp farming in Europe serves as an example.

Semi-intensive pond management is the most widespread, and here fish receive supplementary feeding. This type of management is illustrated by the practice of carp farming in South-East Asia.

In intensive fish farming all feed is supplied artificially and the management of oxygen levels in the water is the predominant factor. This is the first step towards industrial production methods offering several advantages. Hygiene can be improved because cleaning is easier, while water, oxygen and the feed can be controlled by using mechanical or automated equipment.

Controlled environment aquaculture aims to be independent of natural conditions and hazards. Nutrition is by either wet or dry feed, which is specifically formulated. The quality of the water is controlled by oxygenators. Where possible the water temperature is optimized — for instance by waste heat. In addition to lined ponds and concrete channels there are also tanks served by recirculation units. With this closed cycle system complete control over the production process can be achieved. Temperature fluctuations can be eliminated. Impurities, toxins and detritus can be filtered out. Although technically complex, recirculation tanks may in the long run become of major importance in the production of fish meat. The process will be accelerated by the deterioration of the quality of natural waters in the temperate regions and because of water shortage in the arid areas.

NET CAGE HUSBANDRY

Large artificially created waters such as reservoirs and quarries cannot be drained or managed like ponds. Apart from traditional methods they can be advantageously exploited through the use of net cages. Floating cages are of importance in areas flooded behind dams in which obstructions, such as timber, are left intact thus precluding fishing with nets. The Volta dam in Ghana is an example.

DUAL-PURPOSE USE OF WATER AND LAND

It is often possible to use areas for both the farming of fish and agricultural crops. Near Munich the sediment ponds of a sewage farm are utilized in this way and the use of rice paddies in Asia for dual cereal and fish production is well established. In both cases there are risks to the fish as the water may be polluted or carry disease. In paddies the use of chemicals for plant protection has been known to cause problems in fish (e.g. *Mugil cephalus*, the striped mullet, in Taiwan).

3. AQUACULTURE — IN PRACTICE

By definition aquaculture is concerned with all useful aquatic organisms, both flora and fauna. So far only a small number of species is being exploited. However, work is in progress at a large number of research stations to ascertain the possibilities for aquaculture of a range of species.

In this book it is only possible to deal with the major representatives of the algae, molluscs, crustacean and fish species which currently are used in commercial aquaculture.

Apart from the sources quoted, this chapter has drawn freely on the work of Bardach, Ryther and McLarney (1972), Koske *et al.* (1973) and Meske and Naegel (1976).

ALGAE AND SEAWEEDS

ALGAE

During recent years the production of unicellular algae has become of some importance in several countries. Production is relatively easy: green algae require only light, carbon dioxide and certain mineral additives. Agricultural and domestic effluent purification depends on unicellular algae living symbiotically with bacteria.

Alga culture, especially of *Chlorella*, has been developed in the USA, Japan, France and Russia. In Germany Kraut and Meffert (1966) as well as Soeder (1969) have done basic work on the industrial production of the unicellular green alga *Scenedesmus obliquus*. This species has a high crude protein content of 50—59% (dry matter) compared with soya at 34—40%. Feeding these algae to fish, rats and pigs has given good results (see below). *Scenedesmus* is also useful in human nutrition and according to Soeder (1969) can be regarded as a new food source which was found suitable for the treatment of malnutrition in children (Gross *et al.*, 1978). In human nutrition the daily intake should not exceed 50 g because of its high nucleic acid content (Pabst, 1975). Production needs to be located in selected areas to avoid contamination by industrial pollutants.

The production of algal–bacterial biomass in effluent has been described by Shelef *et al.* (1978). The climatic conditions in Israel make the use of green algae in the purification of effluent possible and the resulting biomass can be utilized in poultry and fish feeds.

It is likely that blue-green alga of the genus *Spirulina* will have an important role in the future production of protein-rich biomass. These algae are indigenous to Lake Chad and Mexico and have a protein content of 65%. It lends itself to industrial photosynthesis production as in fact currently undertaken in Italy. At water temperatures of 30–35°C the annual yield is calculated to 50 t of algae per hectare. Summaries on the current practice in micro algae mass production have been published by Soeder (1980) and Richmond and Preiss (1980).

SEAWEED

Seaweed has been cultivated for a considerable time but its specialized mass production is restricted to the Far East where red algae (*Porphyra*), green algae (*Monostroma*) and brown algae (*Undaria, Laminaria*) are produced. The protein-rich seaweeds such as *Porphyra* with a 36% protein content are grown for human consumption, as animal feed and manure. Industrially they are processed into alginates and polysaccharides (Bardach, Ryther and McLarney, 1972).

Seventy per cent of red algae produced in Japan is artificially propagated following the discovery of the reproduction cycle by the British botanist K.M. Drew in 1949. These algae are grown on strings or racks of synthetic twine and on nets fixed to posts. The design and construction of these devices varies from place to place.

In Japan 800 kg/ha a year of *Undaria* are harvested and *Porphyra* produces 750 kg/ha in 6–8 months, both as dry matter. It is reported that an annual yield of 2.4 t/ha of the brown seaweed *Laminaria japonica* is achieved with fertilizer in China. In 1975 over 1 Mt of seaweed were harvested in the Far East, Japan providing half the crop.

MOLLUSCS

This section concerns the cultivation of marine molluscs such as oysters and mussels in their natural environment. In the USA and Japan oysters are produced on a considerable scale. In Europe the husbandry of mussels in the bay of Vigo, Spain, is of economic importance. Meixner (1977) has reported on the successful culturing of oysters on the German North Sea and Baltic coasts. Nearly 1 Mt of molluscs are produced worldwide, of which 230 000 t were Japanese oysters.

The oyster larvae are encouraged to attach themselves to a great variety of devices where they remain until they become marketable. The old method of seabed culture is being replaced by the use of collectors as substrate to which the metamorphosed larvae or spat attach themselves. There are great numbers of collector types ranging from bamboo slivers and rope to scallop shells strung on a wire. The choice of collector is based on the location and species concerned. The essential condition for growing oysters is the presence of sufficient phytoplankton. Oysters and mussels are very susceptible to contamination.

In France, given suitable conditions, the oysters *Crassostrea angulata* and *C. gigas* grow to marketable size in 3 years. In Japan *C. gigas* takes 12–18 months. The edible mussel *Mytilus edulis* needs less than a year at 14–16°C. The yield in terms of the edible parts of the cultivated mussel is often extremely high: in the Philippines it reaches 123 t/ha a year, in Spain up to 300 t/ha, while the Japanese oyster produces 33 t/ha (Honma, 1980). In Germany, experiments on the artificial propagation and rearing of the commercially interesting oyster *Crassostrea gigas* have been carried out successfully (Neudecker, 1978).

CRUSTACEANS

Large-scale production of shrimps and prawns is carried on in Asia where, according to Pillay (1976) 15 000 t are harvested annually. India, Indonesia, the Philippines, Thailand and Japan are the main centres. On the coast of Thailand there is intensive cultivation of shrimp of the genera *Penaeus* and *Metapenaeus* over an area of 7000 ha. In Japan over 1000 t per year of *Penaeus japonicus* are produced by intensive husbandry methods.

The shrimps are kept in ponds, in 10 x 100 m concrete ditches, and also in wooden 50 x 10 m tanks with a fairly fast sea water flow. They are fed minced fish, molluscs or small shrimps. To prevent cannibalism ample feed must be provided. Food conversion is rather poor but offset by the high price achieved for the end-product.

Cultured shrimps were first artificially grown in 1934 and since then methods for their optimal rearing have been developed. The larvae are kept in tanks and the best results are obtained at a temperature of 28–30°C and a salinity of 32–35‰. During the various stages of development the larvae are usually given farm-produced feeds, among which diatoms and *Artemia* are important. In the Philippines and Thailand shrimps are often produced in ponds together with milk fish or bandeng (*Chanos chanos*). *Penaeus monodon*, the sugpo prawn, whose larvae are caught to be reared in ponds of special construction, is of some importance. It reaches a weight of nearly 100 g in 1 year. Culturing lobsters (*Homarus* spp.) has not yet proved successful. Most of the experimental work is being carried out in the USA, but like all lobster culturing it is severely affected by cannibalism.

Fresh water crustaceans are cultured in Asia, Europe and the USA but with a success not comparable to the culturing of other aquatic animals. The pronounced incidence of cannibalism among crustaceans following moulting presents considerable difficulties to their commercial exploitation. They represent a luxury food item and the amounts produced are insignificant compared to the output of protein-rich basic foods.

In the Indo-Pacific region the fresh water shrimp *Macrobrachium rosenbergii* is produced in ponds. This species grows to a length of up to 25 cm and is often cultured with, for example, vegetarian cyprinids, milk fish and gourami (*Osphronemus goramy*) (polyculture). As with other crustacea there is also the problem of cannibalism among *Macrobrachium*. Following ecdysis or moulting each female produces about 50 000 larvae which are kept in rearing tanks with a salinity of between 12 and 14‰ at 26–28°C. They are initially fed with brine shrimp (*Artemia salina*) larvae and later with fish or molluscs.

The fresh water crayfish cultivated in North America are of the genera *Procamburus* and *Pacifastacus*. In Europe, particularly in Scandinavia and now also in the United Kingdom, *Pacifastacus* and *Astacus* (which is now renamed *Austropotomobius*) species are cultivated commercially.

MARINE FISH

In the bays of Japan large numbers of sea fish species are raised and grown in floating cages. This makes rational feeding and harvesting possible. One example is the yellowtail (*Seriola quinqueradiata*) whose larvae are caught because to date artificial propagation is only possible experimentally. They are reared in floating net cages or net-fenced sheltered parts of bays. The yellowtail grows from 50 g to 1 kg in 12 months when fed a diet containing at least 70% fish meal. It can reach 5—7 kg after 2 years. The prerequisite is a water temperature of at least 20°C. Optimal salinity is taken as 16‰. Yellowtail farming is undertaken in over 1300 enterprises and annual output has risen from 300 t in 1958 to over 80 000 t in 1973 and 122 000 t in 1978. Both the yellowtail and the red sea bream (*Chrysophrys major*) are now increasingly cultivated in the cooling waters of power stations (Tanaka, 1976). Some fresh water salmonids can be kept in marine cages provided the salinity is similar to that of brackish water, as for example in the Western Baltic.

Experiments in keeping rainbow trout in net cages along the German Baltic coast — salinity 12–18‰ — have been impaired by below 0°C temperatures and by the considerable fouling of the nets by mussels during the summer. However when the environment is favourable good growth has sometimes been observed for trout in net cages. According to Pohlhausen (1978) this is due to a plentiful supply of naturally occurring food such as *Mysis* and *Poreira*. It is thought that in the Baltic net caged salmonids can be kept on mussels as well as on trout feed (von Thielen and Grave, 1976).

The aquaculturing of marine flatfish — plaice (*Pleuronectes platessa*), flounder (*Platichthys flesus*), turbot (*Scophthalmus* or *Psetta maximus*), halibut (*Hippoglossus*), sole (*Solea solea*) and dab (*Limanda limanda*) is being pursued in Great Britain based on the work by Purdom (1972). The commercial production of turbot and sole in particular is successful but the halibut has been abandoned because it proved difficult. Purdom has also achieved increased yields through genetic improvement. Work is also proceeding at nuclear power stations of the British Central Electricity Generating Board at Hinkley Point, Wylfa and Hunterston.

FRESH AND BRACKISH WARM WATER FISH

THE CARP (*Cyprinus carpio*)

The carp is a native of South-East Asia and China. It is said to have been domesticated for 2000 years. It is also of major importance in Japan, the Philippines and India. It has been farmed in central and northern Europe from the twelfth century onwards, where it has since become the most important farmed species. For instance carp consumption in the Federal German Republic exceeded 7630 t in 1975, valued at over £5 million ($10 million), of which 3900 t were home-produced and 3730 t imported from eastern Europe. In eastern Europe — particularly in Russia, Poland, Czechoslovakia, Hungary and Yugoslavia – carp production is a major industry. In Israel it is the most important farmed fish. The carp was introduced to North America in the nineteenth century but although it is angled for it has not developed into a significantly farmed fish. In Africa the carp has so far scarcely assumed any significance. In South Africa pond carp is being studied to assess its potential under local conditions (Bok, 1980).

Traditionally carp is produced in ponds. The type and design of pond varies from country to country, each having developed specialized ponds for the various stages of the production cycle or end-product, such as breeding, spawning, nursery, rearing and wintering ponds. In South-East Asia and the Far East high yields are achieved in rice paddies and drainage channels. Increasingly the carp is produced by intensive husbandry methods in spring- or river-fed ponds (Japan), in net enclosures or in cages (Japan, Indonesia) and in warm water installations (Russia, East Germany, Hungary).

With extensive methods, feeding depends on the naturally occurring pond fauna and flora augmented by manuring. Under semi-intensive regimes the carp receives additional feed which is mainly of plant origin. In intensive fish farming, as practised in the high water flow rate ponds of Japan and in the technologically advanced enterprises of Israel and Europe, feeding is based entirely on specifically formulated concentrates or compounds. For example in Yugoslavia carp output per year is 2 t/ha with manuring and 780 kg/ha without manuring. In Japan up to 100 kg/m^3 of water is achieved a year in high flow rate ponds.

In temperate and subtropical climates the carp is generally propagated in special spawning ponds. In the tropics the fish tends to spawn spontaneously throughout the year. In Europe spawning is commonly artificially induced by hormone treatment in warm water plants.

The growth potential of the carp is controlled by water temperature. Depending how well they are fed individual carp increase in weight by up to 2000 g a year in tropical zones though they are commonly marketed at 500 g. In Central Europe it takes the carp 3 years on average to reach its marketable weight of around 1.5 kg. In warm water culture the carp's growth parallels that of the tropics (Meske, 1973).

As in the past the carp will continue to be a commercially important fish. The species is omnivorous and under traditional methods of management it makes near-optimal use of naturally occurring pond food, subject to the incidence of insolation and manuring. When additional feed is provided the carp can utilize a wide range of by-products and offal, both from vegetable and animal sources. It has been called the swimming pig. This response to the large spectrum of feedstuffs is used experimentally for the formulation of concentrates (see Chapter 5).

HERBIVOROUS CYPRINIDS

The main species of herbivorous cyprinids are:

grass carp   (*Ctenopharyngodon idella*)
silver carp   (*Hypophthalmichthys molitrix*)
big head   (*Aristichthys nobilis*)

These fish are natives of China while some commercially less important species can be found in South-East Asia. It is likely that after centuries of cultivation in China these cyprinids were introduced into Japan and South-East Asia. During the last 30 years they have spread initially to Russia and later to Hungary and the rest of eastern Europe. They have also assumed some importance in Israel. Interest is now being shown in the grass carp in western Europe.

In China polyculture is the established standard method of production in which all pond organisms are utilized. In addition to the three species listed above the mud carp (*Cirrhina molitorella*) and the black carp (*Mylopharingodon piceus*) are also used in polyculture. The natural productivity of ponds is raised by manuring, as for instance keeping pigs on slats over the pond, and the varied pond fauna and flora is exploited by the different species. The silver carp feeds on the phytoplankton, the big head on the zooplankton, the grass carp on the higher aquatic plants, the common and mud carp on insect larvae, and the black carp on snails and molluscs. Further the faeces of the grass carp are eaten by the common and the mud carp or act as pond manure. Milk fish (*Chanos chanos*) and mullet (*Mugil* spp.) are sometimes included in multi-species populations. Production in these ponds can reach 3.5 t/ha.

These polycultured plant-eaters are not normally given additional feed. When they are it consists of vegetable matter. Grass carp have been kept on pelleted concentrates for years experimentally (see Chapter 6).

In China this fish spawns naturally once a year. In the tropics, as for example in South-East Asia, natural spawning is irregular. As a consequence hypophysation is widely practised. The grass carp needs special conditions such as fast-flowing warm water to keep the eggs in suspension.

The grass carp achieves the highest weights among the herbivorous cyprinids; 10—15-year-old specimens from Chinese rivers weighing up to 50 kg have been reported. In Israel they reach 2.5 kg in 180 days. In South-East Asia the market weights are considerably lower. In China the silver carp is said to reach 20 kg in 12—15 years and the big head can get to 35—40 kg. In Hungary 4—5-year-old fish at 10—12 kg are known.

These herbivores are all of major importance to fish farming in countries enjoying a warm climate. The grass carp converts plants directly into valuable protein although its food conversion efficiency is very moderate (Stott, 1979). This has led to a comparison of this fish with grazing livestock.

The grass carp is likely to be of considerable interest to African aquaculture because of the enormous amounts of aquatic plants, such as the water hyacinth (*Eichornia crassipes*), available in some regions. However thorough investigations are necessary to establish that these are acceptable food and whether the environment is suitable for uncontrolled multiplication and some reservations have been noted by Stott (1979). At Pietermaritzburg, South Africa, artificial propagation of the grass carp is now successfully carried out (see Chapter 6) and based on the conditions and experience in South-East Asia it is likely that this development may be adopted for tropical areas with warm climates.

TILAPIA SPP.

The cichlids generally called tilapia have recently been re-classified into two groups according to their method of breeding. The mouth breeders now belong to the genus *Serotherodon* while the substrate breeders remain with the genus *Tilapia*. Here the common usage term tilapia is retained. The most important tilapia species are:

Tilapia mossambica	the Java tilapia	(now *Oreochromis mossambicus*)
Tilapia nilotica	the Nile tilapia	(now *Oreochromis niloticus*)
Tilapia aurea	the golden tilapia	(now *Oreochromis aureus*)
Tilapia galilaea		(now *Sarotherodon galilaeus*)

The numerous tilapia species are natives of Africa and the Near East. It is taken to be the fish of the bible and a record has been found of it on an Egyptian tomb. Apart from the carp, tilapia is the most widely cultivated fish in the world: *T. zillii*, *T. mossambica*, *T. hornorum* and *T. guinensis* are farmed in their native Africa and the Near East, *T. nilotica* and *T. aurea* in Israel. Many other species are found in wild waters, on fish farms and at experimental installations. During the last few decades several tilapia species have been introduced into a number of countries where several of them have become of importance to aquaculture. An example is *T. mossambica* in Indonesia, The Philippines, Taiwan and Malaysia, while experimental work is in progress in America. In South America, in Ecuador for instance, it has been recognized as a useful fish to farm. Experiments have shown that it is possible to culture this fish in warm water plants in Europe.

The common practice is to farm tilapia in ponds. The pond size and design varies with regions. A number of tilapia species tolerate salinity up to marine levels and therefore they are also raised in brackish water (e.g. the Philippines). However, the majority are farmed in fresh water in enterprises ranging from East African family ponds seldom exceeding 40 m^2 to large lakes. In the large ponds of China herbivorous tilapia species are polycultured with milk fish and mullet. In Israel they are kept together with mullet and silver carp while in Taiwan they are grown with carp only. The performance of this fish in net enclosures is being experimentally evaluated, *T. nilotica* in the USA and *T. mossambica* in Guatemala.

For the most part tilapia are herbivorous. Some species, such as *T. zillii* and *T. melanopleura*, prefer higher plants but the majority, such as *T. nilotica*, *T. mossambica* and *T. machrochir*, feed on phytoplankton. Nevertheless they are capable of utilizing food of animal origin. Other species, such as *T. sparmanni* and *T. esculenta*, are omnivores.

For the nutrition of tilapia an abundance of plankton, algae and other pond organisms is essential. It is necessary to stimulate their growth by manuring. Inorganic manuring, e.g. superphosphate in South Africa, but more often organic manuring, such as domestic effluent in South-East Asia and green stuff in Indonesia, is used. Vegetable by-products are used when tilapia are given supplementary food. Examples are rice bran, broken rice, oil cake, rotten fruit, kitchen refuse, etc. These in turn also contribute to the manuring, giving additional stimulation to pond food production. There are no data on the feed requirements of most tilapia species. Information in this important basic field is of course essential if a rational exploitation of these fish is to be carried on. At present the subject is being researched at Alabama, USA and in Tel Aviv, Israel. In Singapore yields of 1700 kg/ha are achieved in tilapia ponds but bigger crops have been reported.

Tilapia have two breeding methods. One group are substrate breeders, making nests on pond bottoms or depositing their eggs on smooth surfaces and stones. Examples are *T. zillii*, *T. sparmanni* and *T. esculenta*. The other group are mouth breeders (now genus *Oreochromis* — see above). In *T. mombassica* and *T. nilotica* the eggs are hatched and the larvae retained in the females' mouth. In *T. heudelotii* the male incubates the eggs and in *T. galilaea* both sexes share in the mouth breeding. Unlike other fish the tilapia's problem is its fecundity. The difficulty is to prevent it from spawning as uncontrolled breeding complicates management. Ponds will contain too many fish of different sizes and small unmarketable fish will predominate. Prevention of this embarrassing abundance can be achieved through monosex culture. Monosex culture entails time-consuming sexing and the success rate of 80—90% is not sufficiently foolproof. Interspecific hybridization of several species, such as *T. nilotica* x *T. aurea*, has produced

100% male eggs. However this attempt at monosex culture has as yet found little acceptance in practice.

The growth of tilapia varies quite considerably with stocking density, food supply and degree of salinity. Most tilapia species can reach weights of 600—800 g. These weights can be achieved in their first year if good conditions prevail, and specimens more than 1 kg are known. Because of the lack of technology in reducing the ease with which tilapia breed these weights are not normally achieved. In East Africa 250 g fish are often harvested. The male tilapia grows faster than the female. Hybridized male monosex populations can grow from the egg stage to a marketable 500 g in 3 months, partly because of the added advantage derived from hybrid vigour.

The importance of tilapia farming is likely to increase in the future. Tilapia culture is still a fairly new industry and data on husbandry and feeding are as yet incomplete. Further research and development work will contribute considerably by optimizing tilapia culture in the various countries and water conditions of the world.

MILK FISH (*Chanos chanos*)

The milk fish is found in the warm coastal waters of the Red Sea, the oceans stretching from South-East Asia to south Australia and the pacific shores of Asia, Australia and America (California and Mexico). The milk fish is of commercial value mainly in Indonesia, the Philippines and Taiwan. They are cultivated to a lesser extent in Thailand and Vietnam.

As it has not yet been possible to spawn milk fish in captivity its culturing depends on captured fry. However work on this fish is proceeding, and Chaudhry *et al.* (1978) was succesful in fertilizing milk fish eggs artificially and keeping the larvae alive for a few days. Milk fish spawn in the coastal seas and the fry are netted or collected in specially constructed coastal ponds and tidal creek installations. The fish are reared in brackish water and may later be transferred to fresh water ponds. In Java and the Philippines ponds specific to each age group are used. The ponds must be drainable, free from flooding, on clay and an all-year-round water flow must be assured.

The decisive factor in milk fish production is a pond's capacity to support phytoplankton. This is boosted by the use of green manure with mangroves and rice straw. Pig faeces are also used. Manuring with inorganics is increasing slowly.

The fish live on planktonic algae and detritus. The various growth stages feed on different kinds of algae, starting with blue algae followed by green algae. In the production ponds additional plant matter is often fed, including dried water hyacinth (*Eichornia*) and *Haliphila* and even the red algae *Gracilaria*. In Taiwan the maximum yields achieved are 2.5 t/ha.

Milk fish are often grown in polyculture. In the Philippines, for example, Java tilapia (*T. mossambica*) and the hito or walking Australasian catfish (*Clarias batrachus*) are used. In Taiwan the striped mullet (*Mugil cephalus*) is employed, while in Indonesia shrimps and prawns are utilized. At Lagune de Bay in the Philippines a specialized form of milk fish production is practised. Here these fish are kept in pens of bamboo fencing without any additional feeding.

Pond-cultured milk fish grow up to 800 g in the first year, up to 2 kg in the second and over 3 kg in the third year depending on geographical location and management. In Taiwan there are only 4 growing months and overwintering milk fish are expected to lose 10—17% of their weight.

In principle the production of milk fish is possible wherever its fry can be caught in coastal waters and where geographical locations allow the siting of brackish water ponds. Yet even so it can fail, as for instance in Kenya where no-one would eat an unfamiliar kind of fish. To spread milk fish farming beyond those coastal regions where it occurs naturally is as yet prevented by the inability to propagate it artificially. Once this problem is solved the milk fish could become a valuable contributor to aquaculture, even perhaps on the West African coast.

MULLET (*Mugil* spp.)

The most important mullet species are:

Mugil cephalus	the striped mullet
Mugil capito	
Mugil auratus	the grey mullet
Mugil dussumieri	
Mugil corsula	a fresh water species

The mullet can be found in the seas of tropical, subtropical and temperate coastal regions. The size of the areas over which specific species are distributed varies. While the mullet prefers the warmer waters it is capable of tolerating a range of water temperatures and salinity, which in the case of the striped mullet (*M. cephalus*) is 3–35°C and 0–35‰ salinity.

Mullet species are cultivated to an appreciable extent only in Asia and the Mediterranean. In Asia mullet production centres on Indonesia (*M. dussumieri, M. engeli, M. tade*), India (*M. cephalus, M. macrolepis, M. dussumieri, M. troschelli*), the Philippines, Hong Kong and Taiwan. In Israel *M. cephalus, M. auratus* and *M. capito* are cultivated and in the Venetian lagoons *M. saliens, M. cephalus* and *M. capito* are farmed.

Experimentally mullet are being tested in a number of countries to establish their suitability for aquaculture. These countries include England (striped mullet *M. cephalus*), Iraq (*M. oligolepis, M. capito, M. cephalus*) Nigeria (*M. falcipinnis, M. grandisquamis*), Russia (grey mullet *M. auratus, M. saliens, M. cephalus*), the USA (*M. cephalus*) and Cuba.

As for milk fish culture, mullet production depends on the stocking of ponds with fry caught in coastal waters. They are then reared in brackish water but the final production stages can take place both in brackish water ponds, as in Indonesia, or in fresh water ponds, as in Israel. Because of the mullet's adaptability all degrees of salinity can be found in production ponds. The mullet is normally polycultured with other table fish, such as the carp and tilapia in Israel or milk fish and silver carp in the Philippines. In India they are kept with shrimps.

When young, mullet are plankton feeders. Later most species also eat all kinds of plant detritus and higher aquatic plants. *M. dussumieri* and *M. troschelli* remain predominantly plankton feeders throughout. Ponds are manured to raise output and sometimes the fish receive a considerable amount of supplementary feed during the later stages of the production cycle. In Hong Kong, for instance, they are given rice and peanut offal.

Controlled spawning by means of hormone injection treatment has been successful (Shehadeh, 1970, 1973). Spawning mullet are fragile and difficult to handle. However, feeding the fish during the early stages after hatching has proved difficult to achieve. Hence mullet farming at present depends on naturally spawned fry being caught in the sea.

Growth of the various mullet species is not uniform and naturally also depends on environmental conditions which include the level of available nutrition. The striped mullet (*M. cephalus*) can reach a weight of 1.2 kg in Taiwan, 550 g in Israel but only 150 g in India at the end of its second year.

Once the problem of propagation and fry feeding is solved the mullet could become an important species for the aquaculture of many countries. The mullet is tolerant of a great range of water temperatures and salinity. Its nutritional requirements are modest and the quality of its meat is good. Therefore this fish could make a significant contribution worldwide to the production of table fish.

CATFISH

The catfish comprise several genera of different families. They are found in America, Australasia and Europe. The most important farmed species are:

Ictalurus punctatus	the channel catfish of North America (Fig. 1)
Ictalurus furcatus	the blue catfish of North America
Silurus glanis	the European catfish, European wels or sheatfish
Clarias batrachus	the walking catfish of Asia and Africa
Pangasius sutchi	the pla swai of Asia

All catfish are natives of countries with warm or temperate climates and are predominantly fresh water fish. Most are carnivorous and when farmed need to be provided with animal protein. *Ictalurus* spp. have their habitat in North America while the silurids are found mainly in Europe. Species of the genera *Clarias* and *Pangasius* inhabit tropical rivers and lakes of South-East Asia and Africa.

The farming of the European wels is mainly restricted to eastern Europe. Some species of *Pangasius* are cultivated in Asia, examples being *P. sutchi* — pla swai — in Thailand and *P. larnaudi* — pla tepo — in Laos. Among the claridae *C. batrachus* — the walking catfish — is farmed in Cambodia and *C. lazera* in Egypt and East Africa, to some extent in ponds.

Following the failure of buffalo fish (*Ictiobus* spp.) farming in the south central United States in the early 1960s, *Ictalurus* spp. were developed to create an industry of major importance. Catfish production rose three-fold between 1966 and 1969, from 10 million kg to 30 million kg. In 1964 there were 1.620 ha of water devoted to catfish culture, in 1980 more than 30.000 ha. While a large share of this is produced in Arkansas, Mississippi and Louisiana the fish is cultured on a commercial scale in 18 states roughly within a broad belt between the Great Lakes and the Gulf of Mexico. The commercial success of the catfish industry is partly due to the existence of processors and chain restaurants.

The predominent species farmed is the channel catfish (*I. punctatus*) followed by the blue catfish (*I. furcatus*). Bardach *et al.* (1972) lists seven species of catfish used to varying degrees and for specific purposes:

channel catfish	*I. punctatus*	the most commonly farmed species
blue catfish	*I. furcatus*	the second most frequently used species
flathead catfish	*Pylodictis olivaris*	piscivorous and cannibalistic
white catfish	*I. catus*	

bullhead catfish		the smallest food fish, used frequently in experimental work
	I. nebulosus	brown bullhead
	I. natalis	yellow bullhead
	I. melas	black bullhead

The management, processing economics and marketing of the cultivated North American catfish are well established and documented and do not need repeating here. The fish are kept in ponds of between 0.4 and 16 ha and lately also in cages and channels. Feeding is largely with dry compound feeds. Breeding takes place in ponds or is controlled in aquaria through hypophysation. The annual output on US catfish farms exceeds 2 t/ha. The harvesting weight averages about 450 g subject to water temperature. In Arkansas, for example, this weight is achieved in 250 days.

The European wels (*Siluris glanis*) has been artificially propagated for some years in several East European countries. The fish is grown in ponds, usually together with carp (e.g. in Hungary, Yugoslavia). It grows very fast when provided with a suitable environment and is fed accordingly. In eastern European ponds individual weights of between 0.5 and 1.5 kg can be attained at the end of the second summer.

In Thailand and other Asian countries *Clarias batrachus*, the walking catfish, and *Clarias macrocephalus* are farmed in ponds which are stocked with wild fry. Alternatively, artificial propagation of brood fish is possible. Young wels live mainly off plankton, encouraged by manuring the ponds with offal from local crops. In production ponds the fish are given supplementary feed comprising rice offal and trash fish. In East Africa the omnivore *Clarias lazera* is reared in ponds together with tilapia spp. In Western Africa *C. senegalensis* and *C. kameruensis* are of some interest.

Some *Pangasius* spp., such as *P. sutchi* and *P. larnaudi*, make a considerable contribution to pond culture in South-East Asia. In part these fish are now artificially propagated by hypophysation. Ponds are intensively manured because the fry feed on zooplankton. The fish continue to be fed on kitchen refuse, boiled rice, bananas, etc. until they are harvested. In Thailand and Cambodia cage production of these fish has developed employing bamboo cages measuring about 4 x 3 x 2 m. When stock density is high — i.e. 20—60 kg/m^2 — trash fish diet is mainly fed. The growth acieved by *P. larnaudi* is similar to that of the wels or sheatfish — about 0.5—1.5 kg at the end of the second summer while the weight of *P. sutchi* can reach 4 kg in that time.

Clarias batrachus tends to be harvested from 145 g upwards. Under the local methods of husbandry this cropping weight is attained three times a year.

The wels or sheatfish is not likely to make a contribution to the aquaculture of countries enjoying warmer climates, especially where the farming of indigenous species is well established. Although most species of wels display rapid growth their feeding would compete with the nutritional demands of populations which to a large extent already suffer a lack of animal protein in their diets. The wels species tend to be voracious and fish still forms a substantial component of their feeds. This is demonstrated by the feeding of diets based on fish meal in the USA and on trash fish in South-East Asia.

EELS (*Anguilla* spp.)

The genus *Anguilla* comprises 15 species, most of which occur on the coasts of India and the western Pacific oceans (Tesch, 1973). As far as aquaculture is concerned only two species are of economic importance:

the European eel	(*Anguilla anguilla*)
the Japanese eel	(*Anguilla japonica*)

Eels are catadromous migrants spending the greater part of their lives in fresh water. Little is known about the eel's natural breeding habits but it is generally assumed that the European eel is derived from parental stock which spawn in the Sargasso Sea and that the larvae take 3 years to migrate with the gulf stream to Europe. The Japanese eel spawns in the Pacific off Japan and then lives in Taiwanese and Japanese fresh waters.

In Europe large shoals or strings of so-called glass eels, each with a weight of about 0.3 g and measuring about 7 cm long, arrive at the mouth of rivers usually in the autumn or spring. After entering the rivers the elvers develop into yellow eels and after a number of years start changing into adult silver eels before returning down the river to the sea, presumably to their spawning grounds in the Sargasso Sea although proof of this is lacking. The European eel occurs naturally throughout Europe but its natural distribution is often curbed by dams or other obstructions. Consequently many rivers and lakes are now stocked with elvers. Elvers of the European eel were sometimes also exported to Japan to be farmed for food fish.

The European eel crop is mostly made up from wild fish caught in natural waters. In Germany, however, this satisfies only 10% of the market and 90% of consumption is imported. Consequently many proposals for the intensification of eel culture have been made in that country (Meyer-Warden, 1967). Koops (1965, 1966, 1973) and Mueller (1964, 1967) tried growing eels of various age groups in ponds. Bohl (1970) also reported experiments on feeding eels in ponds. However all their work was carried out in natural waters and was therefore subject to the limits imposed by that environment. To date intensive eel culture comparable to that of carp farming does not exist in continental Europe, although a start has been made on commercial eel farming in tanks or ponds of fish farms and power stations in England.

In Japan glass eels are caught in dip nets while migrating up rivers and put into floating baskets for a few days before being placed in nursery ponds. The rearing of elvers in heated tanks at 23°C water temperature inside greenhouses is common. Geothermal energy may be utilized for this. The eels are initially fed with tubifex worms and later with fresh and cooked trash fish, silkworm pupae, etc. Increasingly concentrates are being fed. More recently eels have been fed balanced compounds as a paste in baskets suspended just above the water's surface. Given optimum water temperatures throughout the year the eel can reach the local marketing weight of 200 g after 12 months, although in most localities harvesting takes place in the second year. Because of improved production methods and an increasing demand, catches of the Japanese glass eel cannot meet requirements and consequently elvers are imported from Australia, the Philippines, China, Europe and the USA.

One of the main problems of programmed eel production is that there are very considerable growth differences between individuals. The cause for this is not yet established but sex is likely to be a factor. It is also possible that in intensive husbandry systems with high stocking rates psychological influences such as cannibalism and hierarchical dominance may be involved. Production of even-sized eels requires frequent grading of the fish which is labour-intensive and costly.

Artificial propagation of the eel has been experimentally achieved in Japan (Yamamoto and Yamauchi, 1974) but until this method of elver procurement can be used commercially Japan will increasingly depend on imports.

Taiwan is another major centre of pond eel production whose output is increasing. Unlike in Japan the eels in Taiwan are cultured with carp, silver carp, big head carp, mullet and tilapia (*T. mossambica*). Trash fish and concentrates are fed.

In general eels are expensive. In Europe it is considered a delicacy, but in major parts of the world the eel is of little interest (USA) or unknown (South America, Africa). As it has no visible scales it is not an acceptable food in many countries on religious grounds.

So far eel farming is a major branch of aquaculture only in Japan and in Taiwan. It is possible that the eel could assume growing importance worldwide if it becomes practical to propagate it artificially and to formulate cheaper feeds based on waste proteins.

OTHER WARM WATER FISH

AFRICA

Heterotis niloticus

This fish is cultured in ponds of tropical Africa, mainly in Nigeria, Zaire and the Cameroons. *Heterotis niloticus* is omnivorous. It filters plankton but can also be fed animal and vegetable products. Artificial propagation has not yet been successful. The fish can reach a length of 1 m.

NILE PERCH (*Lates niloticus*)

The Nile perch is grown in East and West Africa, Nigeria being its major production centre. The Nile perch is a predator which lives exclusively off fish. It will breed even in small ponds and can grow to a length of up to 1.80 m.

Haplochromis spp., *Hemichromis* spp., *Serranochromis* spp.

Many of these species are cultivated in East and West Africa. The fish are mainly carnivorous, feeding on insect larvae, snails and even small fish. Some are mouth breeders (*Haplochromis, Serranochromis*), others are substrate breeders on stones, etc. *Hemichromis*). They reach weights of 500—800 g.

Labeo spp.

Some of these species are kept in East African ponds. They are omnivorous and match the carp (*Cyprinus carpio*) in size.

ASIA

The rohu	*Labeo rohita*
The fringe-lipped carp	*Labeo fimbriatus*
The canvery carp	*Labeo kontius*
The mrigal	*Cirrhinus mrigala*
The white carp	*Cirrhinus cirrhosa*
The reba	*Cirrhinus reba*
The catla	*Catla catla*

These and other species are cyprinids and are usually called Indian carp. They are polycultured in India in a variety of preferred combinations and ratios. The main concern of Indian fish farmers is to raise productivity of their ponds by manuring, using cow dung, poulty droppings, and compost. Feeding, especially of young fish, with rice offal and oilcake is widespread.

Hypophysation has been used experimentally but is not in commercial practice. Eggs or fry are almost always caught with the aid of special nets and techniques from the wild. They are then transported in earthenware containers to markets where fish farmers buy them for stocking their ponds.

The biggest weight after 1 year is achieved by the catla which reaches up to 4 kg. The mrigal grows to 1.8 kg, the rohu to 1 kg, the fringe-lipped carp to 450 g, the white carp to 330 g and the canvery carp to 300 g in a year.

AYU (*Plecoglossus altivelis*)

The ayu is a relatively small fish which is marketed at 100 g. It is considered a delicacy in Japan where it is cultivated in small ponds of about 15 m^2 and in flow-through channels at 17–25°C. It is a poor converter of food. The ayu is migratory. It spawns in the sea and the fry swim up rivers. These are caught to serve as stock for fish farms. The market weight is reached in under 1 year. Artificial propagation has been successfully carried out, and is now being used on commercial farms.

LABYRINTH FISH

This term includes several genera which are commonly called gourami. The species of most importance to aquaculture is *Osphronemus gouramy*. Spreading from Indonesia it is now cultivated over most of South-East Asia as well as in India and China. Other species produced in ponds are:

The snakeskin or sepat siam	*Trochogaster pectoralis*
The threespot	*Trochogaster tripopterus*
The kissing gourami	*Helestoma temmincki*

After hatching the fish at first feed on zooplankton, then on plant matter. The kissing gourami prefers plankton. Gouramis are also produced after the harvest in rice fields in 30 cm of water. Breeding in ponds occurs naturally, the fish spawning in nests on the pond bottom. Labyrinth fish are mostly farmed in polyculture. Some of the species listed above may be used together, or they are raised with herbivorous carp and tilapia (*T. mossambica*), among other fish species.

SOUTH AMERICA

PIRARUCU (*Arapaima gigas*)

The pirarucu is one of the largest fresh water fish in the world. It is a native of the Amazon Basin and grows up to 4 m long. It is an excellent table fish. Hence it is hunted intensively and can be regarded as an endangered species. For some years now the fish has been bred in the Peruvian upper reaches of the Amazon but farming in ponds is not common. The fish needs animal protein, it must breathe air but it does not survive water temperatures under 14°C. Because of its extraordinary growth potential it has been subject to experimental work at Ahrensburg where growth from 15 g to 1800 g was achieved in 7 months (see Chapter 6).

FISH IN COLDER WATERS

TROUT

The culture of low water temperature fish includes several species of the salmonids. They occupy a major place in fish production of North America and Europe. Of major note are the rainbow trout (*Salmo gairdneri*), the European brown trout (*S. trutta*) and also the brook trout (*Salvelinus fontinalis*) which is grown especially in the United States.

Traditionally trout are produced in earth ponds. These are increasingly augmented and partially replaced by more up-to-date methods using concrete tanks and also net cages in open waters, quarries and the like. The latest development is to keep the trout in vertical silos. This most intensive method requires strict and constant control of water quality and oxygen levels as stock density is very high and has a fish:water ratio of up to 1:3. A supply of pure oxygen is indispensable (Berger, 1977).

The previously common practice of feeding the trout with meat and fish offal is still continuing in coastline enterprises (e.g. in Denmark). However the feeding of dry concentrates alone is widespread. These feeds are readily available from merchants in formulations specific for the various stages of the production cycle and in various forms — as meal, granules and pellets. A considerable amount of research work on the trout's nutrition has been carried out by amongst others Austreng (1978), Beck *et al.* (1977, 1978), Pfeffer and Becker (1977), Steffens and Albrecht (1976), Takeuchi *et al.* (1978) and Tiews *et al.* (1976). In many cases the research was concerned with the substitution of fish meal by other protein feeds.

SALMON

Salmon eggs, especially of *Oncorbynchus* and *Salmo*, are increasingly produced by artificial fertilization. The fry are reared under controlled conditions until they reach the stage when they are big enough to commence the journey to the sea. Management reduces losses to a minimum and the feed is rationed according to need. It is claimed that because of improved husbandry more salmon return to their spawning ground and do so earlier than after natural production.

Salmon farms are now very successful (Purdom, 1982). Many of the farms are still in the development stage (Risa and Skjervold, 1975). Here the production is controlled throughout the cycle from the egg to the point of sale. In the lochs of Scotland salmon are being reared commercially in net cages while in Norway they are kept in the bays of fjords. (Production is now more than 20,000 tons per year.)

4. THE DEVELOPMENT OF NEW TECHNIQUES FOR AQUACULTURE

ENVIRONMENT CONTROLLED WARM WATER AQUACULTURE

Chapter 2 summarized the various methods used in aquaculture, ranging from practices dating back to ancient times to the more recent and technically sophisticated. Chapter 3 surveyed the specific way in which particular species were exploited by practical aquaculture. This chapter deals with the research and development in environment controlled warm water fish farming, which has been carried out at the Ahrensburg Fish Research Institute near Hamburg since 1965. The Ahrensburg system involves high-density stock management and controlled warm water flow. The feeding is almost entirely with commercially formulated concentrates.

Controlling the environment offers significant advantages. The time between generations can be shortened and artificial fertilization, controlled spawning and rearing can make a considerable contribution to efficiency. The application of research work on commercially useful fish has not achieved the same success as has that following research into the husbandry and genetics of other domestic livestock. This gap is basically due to two biological factors which among livestock apply mainly to fish:

1. fish are cold-blooded and therefore more dependent on their environment than warm-blooded animals;
2. with very few exceptions fertilization in fish takes place outside the animal's body.

These factors — which are discussed in detail below — have so far handicapped the expansion of fish farming because they could not be controlled to any appreciable extent. But taking these factors as a base to start with it was possible — during the work on warm water aquaculture — to develop new methods for the management, feeding and breeding of carp. While the carp

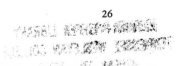

serves to illustrate the potential of warm water culture, other species are equally exploitable, as can be seen by the examples given below.

In Europe the carp is commonly kept in ponds. Under this type of management a number of variable and uncontrollable factors are met, of which the most important is the fluctuating water temperature which in any case is far below optimum in central and northern Europe. During the first winter of life there tend to be high losses among carp due to a number of diseases of varying aetiology involving parasites, bacteria, viruses and nutritional deficiencies. This inhibits constructive rearing and experimental work. The criterion for the work described here was the predictable and controllable rearing of the fish.

STOCK DENSITY*

The biological knowledge on the carp is based mainly on research undertaken in the wild or in simulated natural ponds. Because of the inherent disadvantages of the complex natural system it is very difficult to control the important factors. Appreciation of the shortcomings led to research on carp in aquaria where a controlled environment could be achieved. As early as 1902 Hoffbauer reported that carp kept in aquaria only achieved 10% of the growth of carp in ponds over the same period of time. Langhans and Schreiter (1928) introduced roughly 3-month-old carp into aquaria, but in spite of optimum feeding these fish reached individual weights of only 3–22 g over a 4-year period. They blamed lack of space for the poor growth. After re-introduction into a pond their fish reached 440 g on average within 6 months. This phenomenon was also reported by Walter (1931), Mann (1960), Miaczynski and Rudzinski (1961), with the interpretation that the repression of growth does not inhibit growth potential (Langhans, 1928). Sager (1963) working on goldfish (*Carassius auratus*) observed the same result. Seiler (1938) observed no significant increase in weight of carp in aquaria despite the use of a gravel and charcoal filtering system commonly used for aquaria. In 1963 Krupauer described unsuccessful carp rearing experiments in aquaria. His fingerlings weighed between 0.2 and 0.47 g each after 6 months when the pond-kept controls of the same age had already reached individual weights of 192 g.

Since 1928 the literature referred to the space factor as responsible for the apparent inhibition of growth within the confines of an aquarium. As early as 1924 Goetsch wrote on living space and body size, and Piekarski (1939) commented on the influence of volume on organisms. Langhans and Schreiter (1928) carried out many experiments on the space factor to demonstrate the dependence of growth on space or volume. Willer (1928) tried to divide the space factor concept into excretion, flow and intelligence factors as the cause of growth inhibition in animals kept in small containers. Finally there was differentiation between an absolute and a relative space factor (Lechler, 1934). The space factor complex came to include all these factors and as such was firmly established in the literature (Schaeperclaus, 1961).

---

* Editor's Note: Meske calls this section "Raumfactor" (in inverted commas). The German word means room, space, volume, zone, territory and in this context implies stock density. The choice of word or description caused a lot of problems to the researchers. They used the same word for a number of different aspects. The translation uses the alternatives as applicable. It is assumed that but for the fairly recent political connotations the researchers — and the editor — would have used *Lebensraum* as an apt descriptive term.

To this day the cause of stagnation in the growth of carp and other fish kept in aquaria as described by the above authors has not been fully explained. Experiments at Ahrensburg have clearly shown that the size of a tank in which the fish are kept is not critical, but that the amount of water available is the essential factor. The size of a tank does not stunt fish growth, and this is described in detail below. As to the question of the presence of specific skin excretions containing growth inhibitors, their effect on fish in still water of small tanks needs further examination. From the work by Rose and Rose (1965) with snails, tadpoles and fish it is conceivable that growth-inhibiting excretions may also be found in the carp. Fischer and Albert (1971) found a peptide in the skin secretion of the lamprey which among other things caused contractions of rat uteri. The slime of the fish skin also displays bactericidal and fungicidal properties. It also contains a trypsin inhibitor (Reichenbach-Klinke and Reichenbach-Klinke, 1970). Research at Ahrensburg started therefore on the assumption that volume or density did not influence the growth of carp in aquaria as reported by other workers. It was thought that excretions by the fish, whatever their nature, were responsible for the inhibition of growth of fish in small volumes of water. Exhaustive experiments in this field, covering inhibitors, density and water quality, were carried out by Schulze-Wiehenbrauck (1977) with *Cyprinus carpio* and *Tilapia zillii*. The excretions of fish, including those from the gills, had a marked role in the health and growth of fish in densely stocked tanks without water flow.

After several promising initial studies by von Sengbusch and Szablewski three 1½-year-old carp with an average live weight of 140 g were placed in a 40 l plastic tank on 25 November 1964. The tank had a constant water flow at a temperature of about 23°C. The fish were fed regularly with both live feed (tubifex and daphnia) and dry trout concentrate. The results were startling: after 6 months the fish weighed 915 g, 910 g and 815 g respectively (von Sengbusch *et al.*, 1965). The experiment demolished the so-called space factor effect and represents one of the decisive developments in warm water fish farming.

After a year in the tank the above fish weighed 2815 g, 2285 g and 2210 g respectively (i.e. on 29 November 1965). On 16 March 1966 these fish weighed 4600 g, 4180 g and 3580 g. Figure 2 shows one of these carp after 2 years in the tank, at a weight of 7.5 kg.

A second group of 47 6-month-old carp averaging 10 g were put into a plastic tank on 25 November 1964. Kept under the same feeding and temperature regime they reached an average weight of 850 g after 1 year and over 1200 g on 14 March 1966 (von Sengbusch *et al.*, 1967). After a few months both groups were fed exclusively on trout pellets. The growth rate of both groups refuted the theory that growth inhibition is caused by limited space. Figure 3 shows one of the group in a $60 \times 40 \times 20$ cm tank with an effective capacity of 40 l. The total weight of the 29 carp in this tank was 6260 g, which represents a considerable stock density. At the upper right-hand corner of the tank the water inlet can be seen. The outlet is in the bottom right-hand corner. As soon as consistent growth rates were achieved an installation of more than 100 such tanks was constructed, the outlets all being piped into the re-circulation and filtering system (Meske, 1966) (Fig. 4). Following completion of this installation another similar one was built using glass aquaria (Fig. 5).

TEMPERATURE

The groups of carp described above exhibited growth far above that found in open central European waters. Their growth was independent of the seasons and comparable to the growth

Fig. 1: *Ictalurus punctatus* — the channel catfish as an object of the most successful method of aquaculture in North America.

Fig. 2: Male carp, 3½ years old and weighing about 7.5 kg. This carp was kept in a warm water tank for 2 of the 3½ years.

of carp in the tropics. Buschkiel (1932, 1933, 1939) reported in detail on Galician carp introduced into Java, where they reached 350 g liveweight in 3 months. The males were even sexually mature by this time. These observations show that the growth potential of carp is far above the results achieved in the experimental aquaria, and that this fish can display an exceptional growth rate if stimulated by optimum environmental conditions and feeding. In Israel, too, carp are sexually mature at 12 months (Sklower, 1951). Israeli carp introduced into German ponds revert as expected to the performance of the indigenous fish (Keitz, 1963). Carp imported into Venezuela grew from 150 g up to 1700 g in 9 months (Bank, 1967).

Continuous growth during the winter season is the decisive factor in controlled closed-cycle warm water fish farming. During this period fish in open waters of the temperate zones do not normally display any significant biological activity. Warm water fish farming can make use of a number of industrial energy sources such as renewable energy, fossil fuel and waste hot water from power stations to influence the physiology of fish. While most commercial livestock output can be optimized by providing a suitable and usually specific environment, fish respond to the higher temperature of their environment by a faster metabolic rate, increased feed uptake and a more rapid growth rate. However, above a certain point conversion efficiency is reduced. Since fish are cold-blooded their biological activity responds to van't Hoff's principle: raising the temperature by $10°C$ roughly doubles the speed of reaction. In the European wels or sheatfish (*Siluris glanis*) the digestion process accelerates from 87 hours at $10°C$ to 28 hours at $20°C$ and from 49 hours at $15°C$ to 20 hours at $25°C$ (Molnar and Toelg, 1962).

The optimal temperature for most warm water fish lies above $20°C$. Apart from the tropical fresh water species controlled warm water aquaculture is relevant to the following fish of commercial value: carp, grass carp and its relations, sturgeon, eel and several species of wels and catfish. Certain other species can benefit from a small increase in temperature during certain critical stages of their life cycle.

The benefit of faster metabolic rate in response to increased temperature is possible however only within a defined temperature range. This range is dependent on the freezing and coagulation points of cell albumins. The metabolism—temperature relationship is further limited by the environment of which the species is a native (see Precht *et al.*, 1973). Hence eurythermic fish, i.e. those tolerant of a wide range of temperatures, are the most suitable species for warm water fish farming.

The capacity of certain fish to develop antibodies and therefore immunity to pathogens is at its optimum at $22°C$ (Reichenbach-Klinke, 1976). Abdonimal dropsy in carp, a serious and epidemic disease, occurred just as infrequently at Ahrensburg when carp were kept at higher temperatures throughout the year as it does in the climatically advantaged ponds in Israel. Depending on acclimatization the lethal temperature for carp can be as high as $40°C$. At $38°C$ the fish stops feeding (Hamm, 1964). Schmeing-Engberding (1953) established a preferred temperature range of $16-25°C$ for 8-week-old carp. Experiments at Ahrensburg proved that a steady temperature of $23°C$ throughout the year is suitable for carp, eel, wels and grass carp. It is possible that these species have different optima but the range for carp given above indicates that the limits are not narrow. It is not difficult to maintain a water temperature of $23°C$ in larger installations. At higher temperatures the oxygen supply could present difficulties, and this is particularly critical in re-circulating or closed-cycle warm water fish farming systems (a detailed description of the Ahrensburg system is given below). Higher temperatures reduce the dissolved oxygen capacity of water. At $10°C$ it is 10.67 mg O_2/l, falling to 7.48 mg O_2/l at

Fig. 3: 40 l plastic tank containing 29 carp weighing 6260 g.

Fig. 4: View of part of a warm water research installation consisting of 40 l plastic tanks. Water enters the system from the feeder pipe (top) via the vertical plastic branch pipes. The plastic drain pipes run into the main drain (bottom) which passes the used water through clearing and filter basins under the shelves. From there the water is pumped back up into the heating and header tank at the very top of the installation.

30°C. At the same time the increased metabolic rate due to the raised temperature means a greater oxygen demand by the fish. This in turn leads to a higher feed uptake and voiding of excreta. Under these conditions there is the danger of an oxygen deficiency should the supply not be properly controlled.

WATER QUALITY

The amount of water and its quality is of the utmost importance in warm water fish farming. Through-flow aquaria provide a fundamentally different environment from that found in aquaria with static water. Water flow rate being equal, a smaller tank has the advantage over a large tank because its content is replaced at a faster rate. In contrast the aim for static-water aquaria is to have as large a fish to water ratio as possible.

For experiments on the physiology of nutrition, the senses and the behaviour of the fish, and especially for work in breeding and genetics, the through-flow tank is to be preferred to the static-water tank. In static water the environment is variable and continuously changing. It deteriorates because of excretions and the water needs changing from time to time. This sudden change in the environment is detrimental to the fish and does not allow scientific experimentation which requires constant and comparable conditions. In contrast a constant flow of controlled and monitored water from a central source means that all the fish in all the tanks live under the same environmental conditions throughout the experiment provided the quality and the temperature of that water is maintained. One of the essential conditions for scientific research is therefore satisfied.

When setting up a commercial fish farm a choice must be made between a site on which sufficient warm water of suitable quality is constantly available for a through-flow system and a system which bases the production process on the re-circulation or re-cycling system described below. With regard to the through-flow system there are few suitable sites on which warm water is continuously available in the required quantity and quality. In general these sites are adjacent to industrial enterprises such as power stations — particularly nuclear ones — distilleries and dairies where warm water is a waste product. The most important criterion in the use of cooling water is the volume available. On the basis of results obtained at Ahrensburg the water requirement is a flow rate of one litre of water per minute per kilogram of fish. This rough guide is valid for carp over 50 g with water at 23°C and assumes adequate water quality and oxygen. Provided all environmental factors are given proper attention the water flow can be reduced to 0.3 l/min/kg. However this needs an appropriate feeding technique and requires experience plus a keen eye. The essence of controlled warm water fish farming is that the water is kept at a constant flow rate, temperature, quality and is available 24 hours a day throughout 365 days a year. In practice these conditions are met at few locations only. Industry seldom works round the clock and during the night less warm water is available than during the day. Cooling water availability is also reduced during holidays. Under these conditions the activity of a warm water fish farm is limited by the minimum amounts available.

Apart from the absence of a constant supply, cooling water also displays temperature fluctuations. Fluctuations occur particularly in enterprises which draw their water from rivers, as for example power stations. Coolingwater has a temperature of about 10°C higher than the river, that means a significant drop during the winter months when the water is often below the minimum temperature suitable for warm water fish.

Fig. 5: Warmwater research installation of 108 glass aquaria. This installation has a total volume of 22 m^3 and is subdivided into several autonomous circuits. The heating and distributing header tanks can be seen on the top. The dark pipes carry the water supply and drains. The light guttering collects the effluent for delivery into the clearing tanks which are situated under the passages.

Fig. 6: Detailed view of part of the research installation shown in Figure 5. Water inlet pipe on left. Drainage and aeration pipe in centre. The feed hoppers can be seen above the aquaria. They are rotated by a timeclock-operated electric motor, and dispense feed in controlled amounts.

The quality of the water is decisive for all types of fish farming. Industrial cooling water tends to be subject to three risks:

1. The water quality can be impaired due to industrial processes. Impurities can enter the water because of unforeseen defects or human error, and make it unsuitable for keeping fish.
2. The cooling water can be contaminated because of uncontrolled pollution by poison before it reaches the plant. The onset of pollution may be sudden, rendering the water unsuitable for fish farming. This danger exists especially where cooling water is drawn from river water sources. These sources are constantly under threat of sudden pollution.
3. The cooling water may be of an unsuitable quality from the start. As the level of pollution tends to be on the increase in many areas it can be assumed that there are now only a few rivers from which clean water can be obtained for cooling. In central Europe the cooling water from nuclear power stations sited on rivers cannot always be used as a source of warm water for fish farming because the water quality is frequently poor and unsuitable.

TANKS

The shape of fish tanks and materials of construction are determined by the fish to be kept and their management. Small containers are advantageous for experimental work because they enable many experiments to be undertaken simultaneously with several replications. Glass aquaria are often used because they allow constant observation (Fig. 6). Plastic 40 l tanks made of polythene proved initially successful in the Ahrensburg installation. However they became brittle with time and developed a tendency to crack. PVC tanks last longer. In large-scale installations reinforced glass fibre is likely to be the preferred material for tanks. Alternatively concrete tanks could be used.

The management of warm water fish farming involves the control of flowing water. This needs to be taken into account in the design of the tank, whether it is an aquarium used in research work or a concrete tank used in a commercial enterprise. In rectangular tanks attention must be paid to the flow pattern. This is of particular importance in tanks with unequal sides such as artificial channels. The flow of water along the long axis as, for instance in trout hatching trays, is not the ideal for warm water fish farming. Here an oxygen deficiency can result fairly rapidly because with this system there is a high stocking density and an intensive level of feeding with dry concentrated high-protein feeds. The danger arises particularly in containers with a long flow axis: there is a natural progressive depletion of the oxygen content of the water as it gets nearer the drain end. In consequence an uneven supply of oxygen and feed can occur. When conditions allow it the fish tend to gather tightly at the water inlet, thereby in effect reducing the space actually allocated to them.

Channel-type containers should therefore have a number of water inlets along their longer side and not at the heads. Outlets are consequently best placed on the side opposite the inlets.

Square or circular (radial) tanks are to be preferred. Square and rounded tanks should have several inlets around the periphery with a drain in the centre of the floor of the tank. In the case of radial tanks where a number of sections are grouped around a central drain each segment should be fed by several inlets. The narrowing of each tank section to a central outlet point overcomes the problem of oxygen depletion in oblong ponds and channels referred to above (see Fig. 79).

To remove deplenished water and debris a form of suction pipe is recommended which ensures a constant or minimum water level, helps in keeping the water clear and assists in the removal of debris and faeces from the bottom of the tank. Two types of suction pipe systems have proved successful: Steinberg's pipe and the over-flow suction pipe.

Steinberg's pipe (Fig. 7) is basically an inverted U with unequal sides or limbs. Inside the tank the pipe has an inverted funnel-shaped end sited just off the tank's floor. To prevent fish entering the pipe or being sucked out it is enclosed in a mesh basket. The pipe leaves the tank through its wall and continues down to below the level of the tank floor. This causes the suction effect which removes the debris. This section of the pipe can be rigid or flexible. The first elbow inside the tank has a small screened opening to prevent excessive lowering of the water level in the tank. The screening mesh prevents coarse particles of debris entering the system. When the water level in the tank falls below this opening, air enters the pipe and suction ceases until the water level is restored to its original height by the constant water supply. The cycle then repeats itself. In this system the water level fluctuates continuously over a height equivalent to the diameter of the suction pipe. Leaving Steinberg's pipe the water flows into the main drain, either for disposal as effluent or for re-use or re-cycling after biological treatment.

The overflow suction system (Fig. 8) consists of two sleeved pipes of differing diameter mounted vertically through a hole in the floor of the tank. A single pipe would simply drain the overflow without having a cleaning effect.

To exploit suction for removal of debris the system uses two pipes of differing diameter, one sleeved over the other. The outer pipe is longer than the inner pipe and rests on spacers which form a gap at the bottom through which sediment is sucked up to the drain-away over the top of the inner and shorter pipe. Draining is continuous but suction is not as strong as in the Steinberg system. However in tanks with very high stock density a single overflow pipe may be sufficient because the constant movement of the fish tends to keep solids in suspension.

A simple plug-hole sited in the tank bottom with elbows and an upright drain outlet pipe at the tank water level were found unsatisfactory at Ahrensburg. This type of drainage is commonly used in the farming of salmonids. It was found that faeces and waste feed tended to accumulate in the U-bend. This led to fairly quick putrefaction at the high water temperatures at which the work was carried out.

FEEDING

Until recently carp farming methods were extensive. The fish originally lived solely on the natural products of the pond. As management became more intensive supplementary feeding, predominantly of carbohydrates, increased. Today the most common method of feeding carp is based on natural proteins provided by the pond fauna and flora supplemented by cereals. Supplementary feeding is using cereals common in eastern Europe although different cereals are used.* In Yugoslavia, for instance, maize is used, while in other countries it is mainly barley. As

---

*   Editor's Note: The various cereals used in feeds differ significantly in their calorific values. This may affect the product. The energy content of a cereal is not necessarily reflected in its cost and this needs to be watched when substituting on a cost basis.

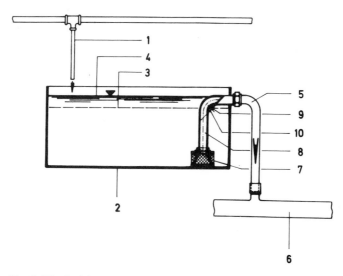

Fig. 7: The Steinberg system. The fish tank (2) is filled by a constant water supply from (1). As the water level (3) rises to water level (4) the water in the pipe (5) starts to drain away (6). This causes suction which removes debris from the tank bottom through screen (7) and pipe arm (8). Flow from the tank exceeds supply, and when the lower water level (3) is reached air enters the pipe through an opening (9) and the outflow stops until the upper level starts the cycle again. A grate (10) prevents clogging up (9). The water level of the tank fluctuates continuously by a depth equivalent to the diameter of the pipe.

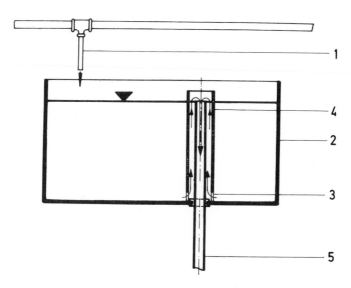

Fig. 8: The overflow suction pipe system. There is a constant flow of water from (1) into the fish tank (2). Water enters overflow pipe (3) at bottom through gap (4) and drains away through the inner pipe (5) carrying debris with it.

stocking density increases protein-rich concentrates are introduced into the feeding regime to compensate or augment the naturally available food in the pond. Von Lukowicz (1976) demonstrated the limits of intensive fish farming in ponds when he neglected naturally available food entirely and stocked the experimental pond beyond its own capacity to remain sweet. The highest yield was nearly 16 t/ha.

Fish can withstand long periods of starvation due to unfavourable environmental conditions or lack of food. It is reported that carp can loose over 25% of their weight in 141—181 winter days when they lack food. Other reports indicate that they die after 159 days. Starvation experiments carried out in conjunction with workers at Goettingen showed that carp lost 40% of their weight in 159 days at water temperature at 23°C. The muscle — or meat — content of the fish in these experiments fell from 45% of total body weight to 33.5% (Pfeffer et al., 1977; Meske and Pfeffer, 1978).

Controlled intensive fish farming requires the feeding of dry concentrates and is more or less dependent on it. This is one of the criteria for this type of husbandry which has already been referred to earlier. This is because natural feed — fauna and flora — is only available during the warmer periods of the year and the provision of natural feed fails in warm water installations because of the problems inherent in their procurement and transport. Dry concentrates have been commercially available for some years in the form of pellets, granules or meal. Now they are also available as floating and non-floating slow sinking forms. However until recently, and at the beginning of the experimental work at Ahrensburg in 1965, the availability of concentrates was restricted to feeds for the salmonids — i.e. trout and salmon. These feeds quickly established themselves for the carnivorous species and became available in a number of formulations suitable for the various stages of the feeding cycle. They replaced the feeding of offal or wet feed — from both fish and the abattoir as a fully balanced single feed. During recent years a wider range of feeds for specific table fish species and their particular requirements at the various growth stages have come on to the market. In Japan, for instance, specific formulations are available for eel, yellowtail (Seriola quinqueradiata), red sea bream (Chrysophrys major) and other species (see Nose, 1978).

FEED QUALITY

It became apparent very early on during the Ahrensburg experiment that carp could also be raised exclusively on dry concentrate feed (Meske, 1966; von Sengbusch et al., 1967). At almost the same time Rajbanshi (1966) and Steffens (1966) reported on early work concerning the use of dry feeds in aquaria. At this stage feeds formulated specifically for carp did not exist, and the early experiments were carried out using trout concentrates. This work proved that carp could be raised using these concentrates. Since then carp feeds have become commercially available although in the main being fed as a supplement to naturally occurring protein-rich food available from the ponds.* Steffens (1979) can be referred to regarding the use of pellets on commercial carp farms.

---

* Editor's Note: In the UK suitable formulations are not always available. Subject to a suitable balance of micro constituents remaining in the feed, high-protein concentrates can be diluted by an appropriate addition of cereals. As the amounts involved are often small, feeds could be mixed manually or the diluting balancer could be fed at predetermined alternative times (Vogt, 1978). Since the cost of a feed is decided by its protein content the lower protein feed may save between 20 and 30% of the cost. As overall the feed cost represents about 60% of the total production cost substantial economic advantages could be achieved even where specific formulations are not readily available to an enterprise.

Feeding exclusively high-protein concentrates to carp in warm water installations produced fish of firm white texture low in oil. The excellent flavour was confirmed in consumer tests. The normal oil content of pond-raised carp lies between 10% and 20%, depending on the age of the fish, the time of year and particularly on the feed. High-carbohydrate diets, as fed in eastern Europe, tend to result in carp with an high oil content. Israeli carp fed sorghum in experimental ponds gave an oil content of between 17% and 20%. In contrast carp fed protein-rich pellets averaged 7%. In some experiments at Ahrensburg extremely low oil contents – as low as 1.3% of liveweight – were achieved.

Controlled environment warm water fish farming offers a number of advantages for influencing the quality of fish by the following means:

1. quality of the feed components;
2. quantity control and frequency of feeding;
3. control of the environment, particularly of water temperature.

Fish meat quality – flavour, texture and colour – is critically determined by these factors and any changes in or between them. The quality of table fish can be tailored to meet specific consumer preferences not only by the feeding programme – in terms of quality and quantity – but also by influencing the metabolism through changes in temperature.

Growth is a function of food – i.e. energy – consumed. Food serves two purposes: maintenance and production. Maintenance sustains the organism's body utilizing in the main carbohydrates and oil. Production or growth requires protein. To increase its weight the body requires a surplus of food or energy above that needed for maintaining itself.

Establishing the effectiveness of the level of the protein content of feeds is important and there are two methods which are increasingly used, particularly in research. There is the protein efficiency ratio (PER) which measures unit growth per unit protein input. The other yardstick is the productive protein value (PPV). This compares the amount of protein in the feed with that in the product.

Prepared feeds and concentrates usually contain vitamin and trace element supplements. Vitamin requirements, particularly for the trout, are well documented (Halver, 1978). However, feedstuffs deteriorate during storage which may be of too long duration or at too warm a temperature, and this may have to be considered when calculating the supplement inclusion rate. The shelf-life of additives or supplements must be taken into consideration as otherwise there may be, for instance, a reduction in the effective vitamin content. In view of possible losses in the nutritional value of a feed it may be advisable to increase the inclusion rate of additives unless stabilized vitamins are used and no imbalance can occur through excessive inclusion. Unless circumstances make it impossible, only fresh food should be used, preferably within 14 days of preparation. It should also be noted that, however good the quality of a feed or its components, the preparatory processes – milling, mixing and pelleting – may adversely affect the feed and its storage life. The influence of trace elements on growth rate is not in dispute (Lall, 1978), and their positive effect on young carp, for example, has been clearly demonstrated during the experiments at Ahrensburg. Where there are trace element deficiencies, reduced or sub-optimal growth results and the effect can also be seen in the incidence of disease. Recent work indicates that fish may be capable of utilizing some trace elements from water (Meske and Pfeffer, 1979). Frenzel and Pfeffer (1982) reported that fish can obtain substantial amounts of Ca, Mg and K from the water. Feeds should not, of course, contain toxic matter. Contamination does, however, sometimes happen and can lead to obscure conditions which may be difficult to diagnose but still cause damage to the fish. In many countries the law imposes upper limits or controls on the presence of the major poisons.

In recent years an acute shortage of fish meal was experienced due to the greatly reduced anchovy catches off the coast of Peru. Although supplies are again improving, a permanent global shortage of fish meal is likely in the future. Shortage also means higher prices. As a result research everywhere is concentrating on finding effective substitutes for fish meal in stock feeds, the aim being to replace all or part of this nutritionally important source of animal protein. Among the substitutes, or rather replacements, which have been tested are ground feathers, poultry offal, alkane-grown yeast, dairy products, whey and powdered dried milk, green algae, krill and bacterial protein among others.

It has been shown that there is a range of possible substitutes for fish meal in trout feeds. These include feedstuffs of vegetable origin not previously considered suitable. Their possible inclusion rates were also found to be higher than was thought likely (Steffens and Albrecht, 1976; Tiews *et al.*, 1978). Experiments at Ahrensburg with carp and grass carp (*Ctenopharyngodon idella*) demonstrated that these fish could be raised without fish meal. It was possible to formulate and test feeds containing only a few basic ingredients which performed as well as feeds containing fish meal. The addition of unicellular green algae produced some interesting results and these are discussed below.

FEED QUANTITY

The application of knowledge on feedstuffs and their optimal formulation is of no avail if feeding methods are not used to the best advantage or when actual feed requirements are not known. Confusion also frequently arises because workers in different disciplines tend to use the same terms to mean quite different things. This section is concerned with intensively managed fish fed concentrates only and it may be useful to define the terms used in this context.

To the nutritionalist a ration means food, any food, or diet. To a farmer a ration is a specific amount of feed dispensed within a given time span. Hence a daily ration or feeding rate of 3% refers to the dispensing of feed equivalent to 3% of the fish's weight per day or a feeding intensity of 3%.

Experience at Ahrensburg indicates that carp on dry concentrates and kept at 23°C consumed a daily ration ranging between 10% and 2%, the amount diminishing with increased weight. A feeding rate table (Table 3) is given below and descriptions of feed utilization experiments are given in the feeding section (p. 94).

The use of a suitably formulated feed combined with an appropriate feeding technique — specifying method, quantity and frequency — can be expected to give a good growth curve.*

---

* Editor's Note: At times there may seem to be a conflict between the interests of academic research and those of commercial fish farming; but this is apparent rather than real because if data are not available commercial enterprises cannot optimize their returns or develop new husbandry techniques. Costs and availability of feedstuffs – feeding accounts for roughly 60% of the production cost – vary almost daily. Fast growth,

TABLE 3.

Feeding schedules at Ahrensburg

Mean weight of fish $\bar{x}$ (g)		Amount of feed as percentage of liveweight per day	Feeding frequency
−	10	10	hourly
10 −	20	8	,,
20 −	30	6	,,
30 −	50	5	,,
50 −	100	4	hourly
100 −	600	3.5	,,
600 −	1000	2.5	,,
over	1000	2	,,

Assessing the value of a feed is complex and conclusions must be drawn on the basis of its effectiveness over a given time. The assumed aim is a fast growth rate.

A measure of the efficiency or suitability of a feed is the Feed Conversion Ratio (FCR). This ratio shows the amount of feed required to achieve a unit weight increase in the product. A FCR of 5 : 1 means that a fish was fed 5 kg of feed resulting in an increase of 1 kg in its weight. As dry weights are not usually used, and assuming that a concentrate contains 10% water, and fish about 70% of water, the FCR can be quoted as being better than 1:1. It is plausible that better feed conversion can be achieved in fish than in warm-blooded animals which have to use a significant part of the energy contained in their food to generate body warmth.

While the FCR is of the utmost importance in the assessment of a feeding regime it should nevertheless not be used in isolation. An excellent conversion ratio may be obtained by reduced feeding but this will depress the growth rate − i.e. increased weight over a specified time span. This does not necessarily lead to an overall food saving.

FEEDING METHODS

Feed frequency is the number of daily feeds given. When concentrates are fed this frequency is of very considerable practical importance. If the time span between feeds is prolonged feed can be wasted because it may be drained off by the water flow, pellets can dissolve and low feed conversion efficiency results. Overfeeding can result in undigested food being excreted. Fish without proper stomachs − as in carp − kept in warm water installations should be fed at least every 2 hours. Table 3 gives the daily feeding schedules used at Ahrensburg.

for instance, is important because it tends to improve meat quality and results in lower cost through, for example, a higher turnover per financial year. Greater turnover reduces capital or fixed costs per produced unit and aids the enterprise's cash flow. New markets may be created simply by making a particular species more widely and regularly available or less costly.

Floating mesh frames have been found very satisfactory for feeding eels with pasty food.

Feed losses must be avoided or at least minimized. Pellets suffer from abrasion or friction, which causes dust. This may be due to the method of processing or subsequent handling. In both cases feed may be lost before it gets to the fish. Dust causes deterioration of water quality. It uses up oxygen and impairs feed conversion. As mentioned above, unsuitable or prolonged storage causes substantial losses. Spoiled or decomposing food inhibits potential growth and may lead to conditions due to vitamin deficiencies and the presence of toxins. Appropriate consistency and size of pellets also needs to be considered. In carp in particular feed utilization declines when the pellets are too hard. The fish at first chew the pellet and then expel part of it from their mouths. Before they take another bite some of the material is lost. Pellet size should be tailored to the size of the fish being fed. For carp pellets which are too small are better than pellets which are too large; they do less harm. Finally attention must be paid when feeding dry concentrates that the feed does not cause oxygen depletion. In the case of warm water table fish part of the feed will remain in the water for a time. The finer the feed particles the more oxygen they will use up. *In vitro* experiments have shown that there is a considerable difference in the oxygen use or loss between various commercially produced concentrates.

A quantity of 20 or 60 g of three different trout feeds were introduced into 3 l glass jars containing tap water. One set was left unstirred while the other was agitated for 16 hours. Both were then analysed for oxygen content. The results are shown in Fig. 9: the differences in O_2 depletion are quite considerable with extreme values of 7, 3 and 0.2 mg O_2/l respectively. The aim should therefore be a regime which ensures that as little of the feed as possible remains in the water.

MECHANICAL FEEDERS

Undesirable wastage of feed can be avoided by mechanical — usually called automatic — self-feeders. These are now widely used, although problems with the lever mechanism are not uncommon with pond carp. For feeding warm water table fish in tanks a demand feeder developed at the Ahrensburg Institute gave the best results. The feeder is operated by the fish tugging at a flexible lure. This feeder is illustrated and described in Fig. 63 while the results of self-feeding experiments are shown in Figs. 64—66. In the event of oxygen deficiency occurring the fish tend to stop using the feeder, thus preventing any further degradation of their environment. Certain feeders dispense a predetermined amount of feed at set times. These may work purely mechanically or they may be electrically operated. There are certain risks involved in the use of these feeders because they dispense feed unrelated to demand and the oxygen state of the tank.

BREEDING

It is possible to influence breeding in warm water fish by appropriate control of their environment. Keeping the water temperature and the lighting constant over the period of a year — or longer — tends to mask the natural seasonal stimuli and the administration of hormones to induce spawning and planned egg production becomes indispensable.

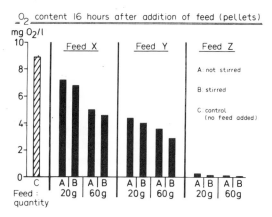

Fig. 9: The effect of dry concentrate feeds on O_2 depletion after 16 hours; (*in vitro* experiment). Hatched column: water only (control); solid column: O_2 content after adding 20 g and 60 g of feeds X, Y and Z respectively; A: without stirring; B: constant stirring; volume of water: 3 l; feeds used: commercial trout pellets (all same size).

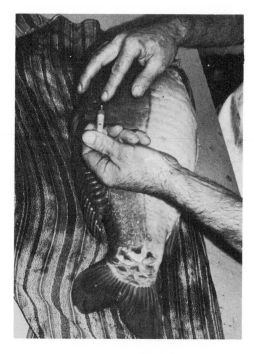

Fig. 10: Hypophysation of a female carp. In this case the fish has been anaesthetized and the injection given in its back. The needle is withdrawn slowly and the area is gently massaged to prevent exudation.

THE BIOLOGY OF REPRODUCTION

In central European ponds the female carp reaches sexual maturity at an average of 4—5 years of age. In males it is 1 year earlier.

With constant temperature-controlled warm water husbandry at Ahrensburg it was possible for fish to attain sexual maturity very much earlier. Female carp produced fertile eggs at 15 months and males were fertile as early as 3 months of age.

In this installation Celikkale (1976) obtained fertile eggs from females at age 11 months and 9 days. Gupta (1975) investigated the histology of gonad development in carp raised from eggs at Ahrensburg at 23°C and described six distinct stages to maturity. The gonadal—somatic index (weight of gonads x 100 ÷ bodyweight) was found to be the same in carp raised in the Ahrensburg warm water aquaria as that of commercially produced carp caged in warm water (Gupta, 1975; Albrecht, 1979). Corresponding experiments with carp in sub-tropical waters were carried out by Bishai *et al.* (1974).

There are two notable benefits arising from an optimal constant environment: the time between generations is shortened and spawning is not dependent on season. Carp do not display endogenous sexual cycles in warm water husbandry. This is in contrast to the carp of European natural waters which spawn only in the spring (Meske *et al.*, 1967). It is not uncommon for carp to fail to spawn at all in northern Europe during cool years, and to spawn only once in years when sufficiently warm weather prevails.

Using the methods described below it was possible to obtain fertile eggs at any time and even during the winter (Meske *et al.*, 1968). Furthermore it is possible to get the same fish to spawn several times during a year. One female produced eggs five times during 1970, altogether amounting to 2.15 million.

Suworow (1948) reported that spawning is temperature-dependent once the fish is sexually mature. After completion of the yolk formation (stage 4) and following a courtship display, spawning commences both in natural waters and in spawning ponds. The display is triggered off by a combination of several environmental factors which act as stimuli to a chain of hormonal activity (Woynarovich, 1953). Among the external influences are rising water temperature, increasing light intensity, the lengthening of days and the presence of aquatic plants in shallow waters — such as flooded meadows — which provide tactile stimulation. These and possibly other factors give rise to a series of endocrine processes, not all of which are as yet clearly understood. This endocrine activity culminates in the secretion of gonadal hormones under the influence of the pituitary gland or hypophysis. This is the last stage of the egg-forming cycle and starts the process of ovulation (Hoar, 1969).

It is not yet established which part of the pituitary plays the decisive role. Sinha (1971) was able to separate fish pituitary extract into three fractions. Of these the second fraction caused only spawning behaviour and ovulation. This fraction contained gonatropic substances.

In warm water fish culture these reactions are influenced by the controlled constant environment. External factors are largely or completely absent and this results in the suppression of the seasonally induced endocrine changes found in nature.

BREEDING TECHNOLOGY

Ihering described artificial spawning in the Characidae as early as 1935. His methods were developed and made practical for carp by Woynarovich (1961, 1964). The process is called hypophysation and consists of intramuscular injection of a suspension or extract of pituitary gland material. This raises the level of sex hormones and brings about the maturation and shedding of the sex products in both male and female fish. Woynarovich used the method he had developed on the carp in the Ahrensburg aquaria in August 1966. The carp were 2 years old when the treatment commenced, and had been fed almost exclusively on dry concentrates (Woynarovich and Kausch, 1967). The result was astonishing because the carp in the experiment was only 2 years old and August is an unusual time for carp to spawn in Europe.

Hypophysation is an indispensable aid to breeding warm water fish culture. The method used by Woynarovich is summarized as follows:

In hypophysation the dosage is determined by the sex and weight of the fish to be induced to spawn. The dosage for a female is one pituitary per kg of fish; that for a male is one pituitary per individual. The pituitaries are pulverized after drying, and dissolved in 0.3 ml saline and 0.2 ml glycerine. The appropriate dosage is slowly injected deep into the musculature of the back* (Fig. 10) after the fish has been anaesthetized for about 2 minutes in 0.006% solution of Sandoz MS 222. The genital opening of the female can be closed by a stitch to prevent premature shedding of the eggs. In general the fish can be stripped about 16 hours after hypophysation, the stitch having been removed and gentle massage being applied to the abdomen (Fig. 11). The eggs are collected in a dry plastic dish (Fig. 12). The milt of the male is obtained either through suction with an aspirator (Fig. 13) or by manual pressure on the abdomen and then applied directly on to the eggs.

A solution of 40 g NaCl and 30 g carbamide in 10 l of water is then poured gradually over the eggs and milt to prevent the eggs sticking together or to the container, and to make them swell. The milt and the eggs are then carefully mixed using a plastic spoon or a goose-feather. Fertilization takes place within a few minutes. Dispersion is important and the solute must continue to be added until the eggs cease extending. After about 1 hour the eggs are rinsed several times in a solution of 10 g tannin in 10 l of water to remove all traces of stickiness in the eggs and prevent adhesion. Following this the eggs are placed in Zuger jars where they are kept in motion by a continuous flow of water from the bottom of the jars. The flow is adjusted to keep the eggs in suspension and in gentle movement without permitting them to be washed out of the jars (Fig. 14). At 22°C cell division commences 1 hour after fertilization and hatching after 62.5 hours (Neudecker, 1976) (Fig. 15).

The pituitaries are obtained by dissection of the head of sexually mature fish. Hypophysation is now used widely in intensive fish farming throughout the world and is applied in the controlled spawning of numerous table fish species (Gerbilski, 1941; Atz and Pickford, 1959, 1964; Anwand, 1963; Steffens, 1957; Mittelstiller and Hamor, 1961; Dadzie, 1970; Bardach et al., 1972, and others). A global review of the literature on this subject is given by Donaldson (1977). Hypophysation as a means of planned spawning and fry production is used on a wide

* Editor's Note: a method using a different dosage injected into the peritoneum is described by Bardach et al. (1972), who also stress that carp hypophysation is species-specific and only carp pituitary can be used. However carp pituitary can be used to induce spawning in other fish species. Commercially prepared pituitary extract is now available from the USA and the other hormones have also been used on ripe fish, including tamoxifen and clomiphene.

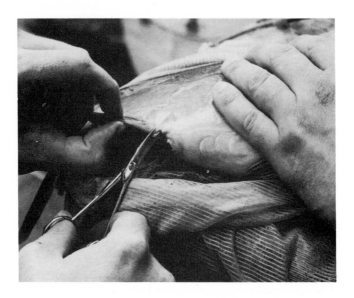

Fig. 11: Removing the stitch from the genital opening of a female carp
immediately prior to stripping.

Fig. 12: Stripping a female carp.

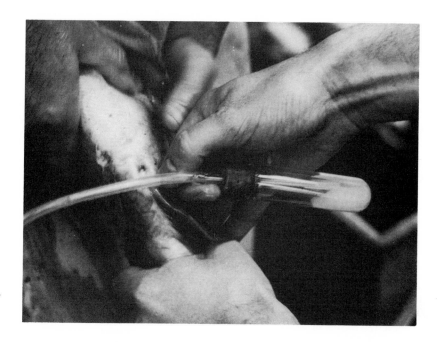

Fig. 13: Extraction of milt by aspirator.

Fig. 14: Zuger jars containing (left to right) 510 K, 460 K, 550 K and 800 K fertilized carp eggs.

range of fish species in eastern Europe, particularly in carp. However full knowledge of the hormonal processes involved is still lacking and consequently it is as yet not possible to establish a standard hypophysation treatment. Nevertheless factors influencing the process can be listed here, although they may not always prevail, be repeatable or be relevant:

1. The sexual and physiological state of the fish receiving the injection.
2. The type, sex and stage of sexual maturity of the donor fish.
3. The value and strength of the gonadal substances in the injected extracts.
4. The environment in which the recipient fish is kept and the husbandry system under which it is managed.
5. The method used to obtain the pituitary, its subsequent handling to point of injection and the season during which it is removed.
6. The method of injection.

For these and other reasons many workers have tried to define the active elements in order to achieve some standardization for induced spawning. Comprehensive series of experiments have been carried out using a wide range of hormones, especially mammalian, without achieving any consensus which would allow specific recommendations to be made. Surveys of results obtained with mammalian gonads, steroids and hormones were presented e.g. by Shehadeh in 1970 and 1973 and by Chaudhuri 1976. Bearing in mind the provisos listed above, the use of fish pituitaries remains the preferred means of hypophysation.

In hypophysation the activation period is water temperature-dependent. Kausch (1975) established a linear relationship for spawning during May in German ponds as $y = 46.78 - 1.407 \, x$, where x represents the temperature of the water and y the time to ovulation in hours.

It is however possible to induce spawning in fish at a constant water temperature without the use of hormone treatment. During experiments untreated mature carp in warm water tanks proceeded to spawn after the introduction of natural turf and artificial lawn (Meske and Cellarius, 1974). It was evident that the perpendicular blades of the grass exerted a tactile stimulus on the fish through the central nervous system, the hypothalamus and the pituitary to cause it to spawn.

Nevertheless hypophysation of farm fish is an indispensable aid to breeding and genetics, particularly so with warm water management. It is a means by which experiments can be conducted into the physiology and preservation of the sexual products. The storage of milt in sperm banks has already been the aim of a considerable body of experimentation and in many species milt storage of up to 1 year at temperatures ranging from $-4.5°C$ to $-196°C$ has been achieved using very diverse media (Shehadeh, 1970; Holtz et al., 1979). The storage of milt made possible the AI (artificial insemination) service in mammalian livestock which itself made an outstanding contribution to animal husbandry.

In mammalian AI only the sperm is stored. In the case of fish the preservation of unfertilized eggs would be an application of major importance in extending the technique. The solution of the fish egg storage problem (Meske, 1968a) would — in conjuction with a milt or sperm bank — give tremendous advantages to breeders and breeding. Carp spawn up to 1 million eggs at a time, and storage would ensure that the production of progeny of genetically improved cock and hen fish could occur long after their deaths. This facilitates back-crossing, the establishing of inbred lines, and progeny testing. All these advantages are of inestimable value in the genetic improvement of stock.

An essential contribution towards the realization of breeding aims and genetic improvement would be a suitable feed for the fry and for rearing them intensively in a controlled environment. This also holds good for out-of-season fish production. The newly hatched carp fry (Fig. 16) first of all attach themselves to any floating or suspended object. For this purpose gauze frames are used at Ahrensburg. Once the yolk is consumed and the fry swims freely it will start to feed on plankton. Initially some difficulty was encountered in the feeding of fry in aquaria but then the larvae of the Californian brine shrimp *Artemia salina* were found to be suitable food and in tests proved to be the best feed. The dried eggs of artemia hatch after 2 days in highly aerated salt water and the resulting nauplii can be kept alive in fresh water aquaria for several hours. The nauplii provide an excellent start for fry because they swim in front of the fish's mouth and are thus readily taken up.

According to Woynarovich and Kausch (1967) fry have a high oxygen requirement and this demand must be given careful consideration: 100 000 freshly hatched carp need 83 mg O_2/h at $20°C$ and 290 mg O_2/h at 10 days of age.

Large quantities of carp fry were successfully raised at high stocking densities in the warm water tanks at Ahrensburg (Fig. 17) (Meske, 1968; Kossman, 1970). The experimental work concerning the feeding of fry is described in Chapter 5, page 67 onwards. The growth advantages of fry hatched and reared in warm water are substantially superior to pond-produced fry. For this reason fish farmers buy these fry every spring to stock their ponds.

BREEDING AND MULTIPLICATION

Breeding and multiplication work is made easier by the exploitation of gains which can be achieved through controlled warm water culture. The methods and observations discussed above open possibilities for the poikilothermic table fish producer which are not available to farmers of warm-blooded stock. A very large number of progeny and a speeding-up of the reproduction cycle can be achieved by influencing the environment. Controlled environment is of course practised to some extent with all domestic livestock but in the case of fish the advantages are very much greater. Eggs and sperm can be produced to a programme at predetermined times and independent of seasons through the use of hormones. Survival and rearing of fry can be made more effective and profitable by controlling the environment, improving husbandry and nutrition using the advantages arising from the experiments described in the next chapter.

THE AHRENSBURG CLOSED-CYCLE SYSTEM

As stated above, constant or near-constant environmental conditions are essential to experimental work on the management, nutrition and breeding of warm water table fish. This also applies in the case of optimal and continuous production in commercial enterprises. As the availability of sufficient water of suitable quality and at the right temperature to run a through-flow system is rare, a closed-cycle system needs to be employed. In a closed system the polluted water draining from the tanks is piped into a purification plant for treatment, and after aeration is returned to the fish tanks. A closed warm water system, suitably housed and insulated, saves energy because heat losses are drastically reduced.

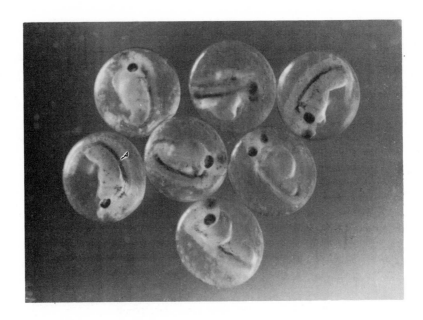

Fig. 15: Carp eggs shortly before hatching.

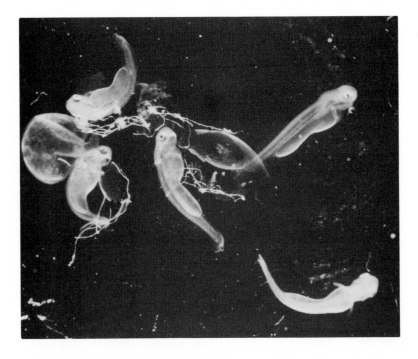

Fig. 16: Carp fry immediately after hatching.

Fig. 17: Eleven-day-old carp fry in an aquarium.

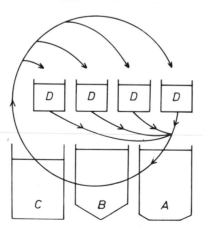

Fig. 18: Water circulation in the Ahrensburg system (schematic — see text).

Closed systems also reduce water requirements — water availability is an increasingly major limiting factor — because in such systems it is only necessary to replace water lost by evaporation, spillage and spray. Furthermore the system is not subject to the natural seasonal temperature fluctuations and it precludes problems arising from the progressive pollution often met in natural waters.

The Ahrensburg system was developed at the Max Planck Institute to make possible a form of aquaculture which would not be beset by the vagaries of climate or the environment (von Sengbusch et al., 1965; Meske, 1967; von Sengbusch et al., 1967). The system was one of the first developed for the production of fish in general and not for experimental work alone. Its basic layout is shown schematically in Fig. 18. Saeki described an unheated closed re-circulation installation at Motokawa (Japan) in 1965 which cleaned polluted water for carp through gravel filter beds. A similar system using gravel filtering was patented by Keely in the USA in 1960. This was designed to be used in the rearing of cold water fish. Parisot reported the development of a system in 1967 which re-circulated the water through a box containing layers of oyster shell, charcoal and gravel. *

The Ahrensburg system was innovative and made a considerable contribution to the then emerging field of fish farming using water re-circulation, mainly because it used no mechanical filter and relied entirely on the biological activity of micro-organisms. Since the time of its inception the system has been repeatedly improved through changes of detail and dimensions. However the basic principles of biological treatment of waste water have remained the same. The system is illustrated in Figs. 19—23, and its construction is detailed in the following description.

CONSTRUCTION

The water from the tanks drains through pipes (1) into the activated sludge (or mud) chamber A. Atmospheric air is pumped into the sludge by a compressor (not shown) through main and feeder pipes (3, 4) into a tubular aerator emitting small bubbles. This supplies the oxygen for aerobic micro-organisms and also creates circular turbulence of the mud, the movement being aided by fillets (5). The fluidized sludge travels between three baffles (6, 7, 8) into the settlement chamber B where the mud settles out, acting as a filter and trapping fine particles at the same time. From the bottom of this chamber (9) the sludge is continuously pumped back into chamber A by three 5 cm diameter air pumps (10). Mud rising to the surface is sucked up and also returned to chamber A (11). The water overflows a partition incorporating a device to exclude mud and flocculates into the reservoir or cistern C (12). Here the clear water is aerated (13, 14) and heated by radiators (15) which are thermostatically controlled (16). The clean water is pumped (17) along pipe (18, 19) back to the fish tanks, thereby completing and repeating the cycle. All the water losses in the system are made good from the mains through a ball cockvalve (20).

---

* Editor's Note: a similar and successful system working on both fresh and salt water fish farms was developed by Cansdale in the United Kingdom during the early 1970s. Yanzito (1981) and Manci and Yanzito (1982) describe a sophisticated circular closed-cycle aquaculture unit containing all stages of fish production using solar or some other renewable energy as heat source. Each unit is designed to produce 20 000 kg a year (see Chapter 7).

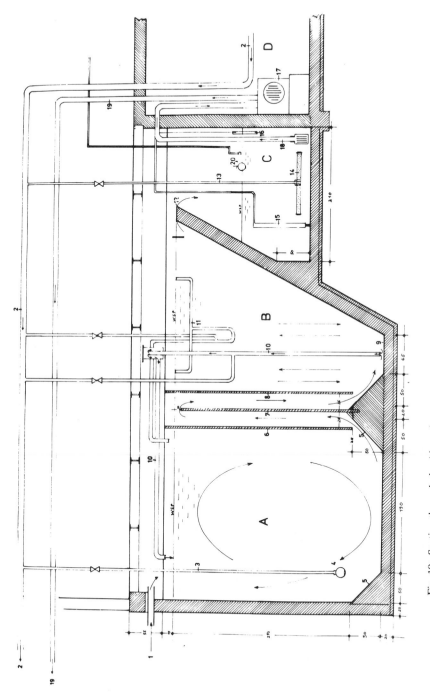

Fig. 19: Section through the Ahrensburg system. Six of these systems are installed; each has a volume of 54 m³.

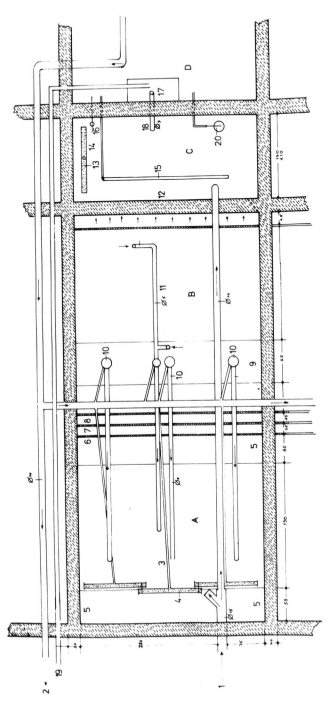

Fig. 20: Plan of the Ahrensburg system.

Fig. 21: View of clearing chamber B, showing the three air-driven pumps which return settled-out sludge to the activated sludge chamber A. The depth of the chamber is 4 m. In the background the inlet bringing the water from the aquaria to the activated sludge chamber A can be seen. At the top are the air supply pipes.

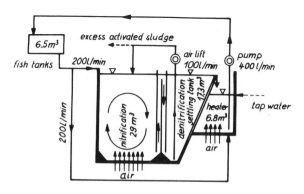

Fig. 22: Ahrensburg system flow rates (Meske and Rakelmann, 1980).

Fig. 23: Water purification plant. Photograph shows six individual units which purify the water in the circuits. Each has a volume of 54 m^3.

Fig. 24: Purification plant. View of the activated mud chamber A. The high rate of aeration can be clearly seen. The air supply lines are on the left.

CLEARING CHAMBER VOLUME AND FLOW RATE

For several years the work at Ahrensburg was carried out with circuits whose total volume was 4 m³ (Meske, 1973). At the same time experience was also gained with 20 m³ systems. Since 1972 60 m³ volume systems, as described above, have been the standard circuits used. The clearing chamber unit consists of a rectangular concrete basin with a depth of 4 m. The activated sludge chamber has a base area of 4 x 3 m and a net volume of 29 m³. The clearing chamber (sedimentation, denitrification) has a floor area of 1 x 3 m and a capacity of 17.3 m³. The angle of the partition wall between B and C is 60°. The baffles between A and B are constructed of compressed asbestos sheeting. All three chambers (A, B and C) have a total volume of 53.1 m³. The rest of the unit, that is the fish tanks within a circuit, contain 6.5 m³ of water so that the ratio of plant volume to tank volume is 9:1. Subject to stock density at any given time, part of the water drained from the fish tanks may be pumped directly into the reservoir C without passing through the previous chambers. Frequently the fish tank water bypasses the treatment chambers at a rate of 200 l/min while simultaneously a further 200 l/min pass through them. The amount of sludge returned from B to A by pumps 10 (Fig. 21) is then in the region of 100 l/min (Figs. 21 and 22). At present four such re-circulation systems are in operation at Ahrensburg to provide the environment for the warm water table fish involved in the experimental work on various aspects of their physiology. Stock density depends on experimental requirements and varies between 300 and 1500 kg of fish — predominantly carp, wels and grass carp — per unit. This gives a ratio of fish to total water volume ranging between 1:200 and 1:30.

HEATING

The circuit water is heated by oil. The water circulating through the fish tanks is heated indirectly by 2 x 1 m stainless-steel radiators containing water at about 80°C. There is one radiator in each unit (Figs. 19 and 20) regulated by a thermostat. This arrangement achieved a near-constant temperature of 23°C in the fish tanks throughout the year.

MATERIALS

It is important to employ the appropriate materials in the construction of an installation as otherwise there may be a build-up of deleterious substances. Copper and copper-containing materials should be avoided. Anything galvanized needs to be treated by a suitable primer and epoxy resin. The use of PVC is recommended for all pipework. Newly poured concrete can cause death and diseases from high pH values due to its alkalinity. Hence concrete chambers must be either thoroughly washed down for prolonged periods or suitably treated. Application of a primer and an epoxy resin coating gave satisfactory results at Ahrensburg (Fig. 23).

OPERATION

To maintain steady growth in the fish, and to keep production at a constant level, it is essential that the flow of water and its temperature are both maintained. Power cuts occur and the electricity supply needs a back-up. A 75 hp diesel engine with an 87 kVA output and automatic cut-in was therefore installed at Ahrensburg to ensure continuity in the vital electricity supply to the system.

STOCK DENSITY

Experience has shown that the installation described here can be stocked in the ratio of 1 kg of fish to 30 kg of water, that is 1:30. It can be presumed that this ratio could be improved further as it is not actually necessary to pass all the water draining from the fish tanks through the purification plant. As has already been mentioned the process can be short-circuited by a by-pass pipe which feeds the drained water straight into chamber C for immediate re-use.

AERATION

The choice of compressor for a plant is determined by a capacity to provide six times the total volume of the plant per hour. A 588 m^3/h unit was therefore installed, and as in the case of the electricity supply a stand-by facility was also provided. A fully installed and operational reserve unit should be considered essential. The activated sludge is aerated through a porous aerator (Fig. 24).

WATER PUMPS

Initially submersible pumps were used to propel the water through the circuits. On the whole they did not prove satisfactory because they tended to develop defects. They leaked both oil and water and were replaced by rotary pumps throughout the plant. Rotary pumps have since given trouble free service.

CRITERIA

The construction of the Ahrensburg system plant is not dissimilar to industrial or domestic sewage works. The function of the Ahrensburg unit, however, cannot be compared with them.

1. In the Ahrensburg system the same water passes and re-circulates continuously through the plant, and the complete purification of the water in one cycle is not necessary. In industrial and domestic installations a high degree of water quality is achieved during one passage through the purification process.

2. Industrial and domestic sewage works are designed to cope with heavy demands. For the biological processes to function properly a minimum load is stipulated as the micro-organism would otherwise not get enough nutrients. If the in-effluent is too dilute or too pure the organisms of the activated sludge would starve. The same applies to a once-through purification plant fed the water from aquaria. During periods of low loading such as occurs during the nightly feeding pause the plant would receive almost pure water. In a re-circulation system the micro-organisms are provided continuously with a feed substrate and one single passage through the system does not remove all of the nitrogenous impurities.

3. The water in the re-circulation or closed system can pass through the clearing chamber faster

than is usual in sewage treatment. The flow rate is determined by the time allowed for the activated sludge to settle out in chamber B. The dimensions of the Ahrensburg plant are significantly reduced compared with a typical sewage works because the time the water spends passing through the system is very much shorter.

4. Biological treatment plants commonly operate in the open. Their efficiency is therefore subject to the prevailing ambient temperature towards which the effluent's temperature changes. It is a criterion of the Ahresnburg system that the water is heated and kept at a thermostatically controlled steady temperature. At the temperature required for warm water fish culture − i.e. $23°C$ − the behaviour of the activated sludge is obviously quite different to that in unheated plants. This applies both to the spectrum of micro-organisms and the rate of their activity. The intensity of micro-organism activity increases as temperature is raised. However, detailed investigation into the protozoan and bacterial component spread in the warm water-activated sludge of the Ahrensburg system still needs to be carried out.

FUNCTION

The activated sludge or mud method is one of the most efficient means of water purification. The sludge contains a substantial concentration of various species of bacteria which to a large extent form colonies. Together with very fine detritic particles they form gelatinous clumps which in the absence of adequate turbulence flocculate into larger crumbs. This flocculation makes the separation of sludge and water in the reservoir possible through sedimentation and is an important factor for the successful operation of the activated sludge system.

The bacteria in the sludge feed on the colloidal and suspended organic matter contained in the effluent. The process primarily involves enzyme and osmotic activity and also the adsorption of pollutants by the sludge crumbs. The bacteria utilize these substances in their metabolism and this process requires oxygen. The spectrum of bacterial species varies in accordance with nutrient availability, oxygen levels and temperature, and is possibly augmented by protozoa and certain metazoa on occasion.

In a closed-water circuit the load on the purification plant consists entirely of wasted feed and fish excreta. Certain heterotrophic bacteria convert the waste and detritus into ammonia. This, together with ammonia produced directly by the fish, is oxidized by autotrophic nitrifying bacteria (*Nitrosomonas spp*) to nitrite:

$$NH_4^+ + 1.5\ O_2 \rightarrow NO_2^- + H_2O + 2H^+.$$

Other bacteria (*Nitrobacter spp*) convert the nitrite to nitrate:

$$NO_2^- + 0.5\ O_2 \rightarrow NO_3^-.$$

Both reactions are endothermic.

The microbial oxidation of ammonia to nitrate is called nitrification. It is the most important factor in the purification of water in a closed system. As the bacteria involved are aerobes the nitrification process requires an adequate supply of oxygen, and this is provided in the active chamber A (Fig. 19) which is constantly aerated.

The opposite occurs in denitrification, which is achieved by anaerobic bacteria in a medium completely or nearly completely devoid of oxygen (sedimentation chamber B, Fig. 20). Here nitrates are reduced first to nitrites and then to nitrogen in the presence of organic carbon compounds:

$$10\,H^+ + 2\,NO_3^- \rightarrow 2\,OH^- + N_2 + 4\,H_2O.$$

The gaseous nitrogen rises in the sediment chamber B, taking large sludge aggregates with it to the surface to be sucked up (11, Fig. 20) for return to the activated sludge chamber A. The processes involved in the denitrification of sedimentary tank sewage works have been described by Viehl (1949), Rueffer (1964) and others.

Figure 25 shows the analytical data on the water from one of the original Ahrensburg closed circuits of 4 m^3 volume, together with the growth curve of fish raised in it. The data relate to a 17-week period and are representative of data obtained over the years.

The big 60 m^3 closed circuits operating to the plant design shown in Figs. 20—24 have now been in use since 1972. The ammonium (NH_4^+), nitrite (NO_2^-) and nitrate (NO_3^-) values have been regularly determined photometrically using a Zeiss P4 photometer. The results obtained from two circuits over a 3-year period were published in 1976 (Naegel et al.). During these 3 years the values for NH_4^+ were less than 1.25 mg/l and for NO_2^- less than 0.5 mg/l. Most of the time the NO_2^- values were less than 0.2 mg/l.

On the basis of 18 years of experience with activated sludge circuits the following observations can be made: nitrification in adequately aerated activated sludge chambers presents no difficulties and regular readings showed that the concentration of ammonia and nitrites exceeded 1 mg/l only on rare and exceptional occasions. As denitrification proceeds in those zones of the plant containing no oxygen high nitrate values — which may reach 1000 mg/l and even peak at 1800 mg/l — can be expected unless a denitrification chamber is incorporated in the design of the system. The loss of water from a circuit through evaporation, spillage and removal of surplus sludge ran to an average of 2% per day of total circuit volume over a period of several years. This prevented the build-up of excessive nitrate levels. It kept the concentration below the limits which could adversely affect carp, wels and grassfish.

Continuous recording of the performance of the fish kept in the system's circuits demonstrated a high and constant growth rate. This steady growth rate was independent of observed variations in the nitrate levels (Fig. 26) and this was confirmed repeatedly throughout the experimental work (Naegel et al., 1976; Meske, 1976b). Knoesche (1969) reported that he was unable to find any damaging effect on carp with nitrate values as high as 3000 mg/l.

In the Ahrensburg system the water in the circuits is continuously polluted by excreta and waste feed. Consequently there is a steady increase in sludge concentration. To keep the amount of sludge in the activating chamber at a constant level surplus sludge should be drawn off fairly frequently, i.e. weekly. The method used at Ahrensburg was to pump the sludge into gauze strainers placed above the plant. The water was allowed to drip back into the system and the sludge was removed from the plant.

SALT WATER MODIFICATION

During recent years modified experimental circuits with a 2.5 m^3 capacity have been built to

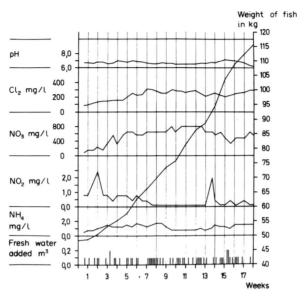

Fig. 25: Water analysis and fish growth data of circuit 601 during an 18-week period.

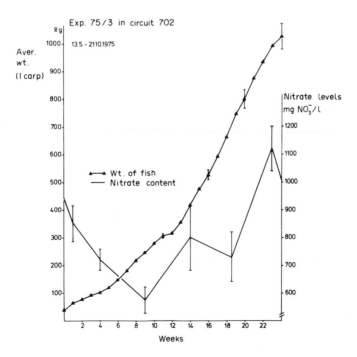

Fig. 26: Levels of nitrate concentration and average growth rate for individual carp in a closed warm water circuit during a 24-week period (Experiment 75/3).

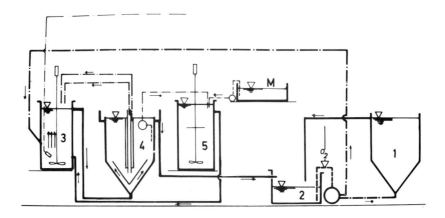

Fig. 27: Experimental salt water closed circuit incorporating special denitrification tank.
For details see text.

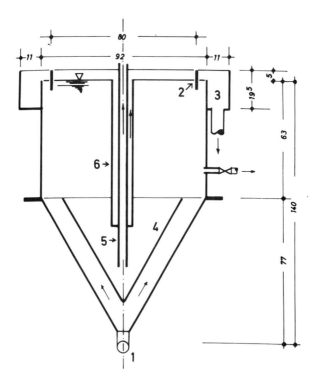

Fig. 28: Detailed section of sedimentation tank of the experimental salt water closed circuit. (See Figure 29 for external view.) Activated sludge and water enters at (1), settles in inner cone (4) and is drawn off through pipe (5) for return into activated sludge tank. Toothed skimmer (2) prevents scum flowing into clear water drain.

test the suitability of the Ahrensburg system for salt water fish culture. Instead of using large concrete chambers the purification of the salt water plant is based on plastic tanks of various shapes and sizes. This makes for greater flexibility both in design and usage. A special additional denitrification tank was slotted into the system between the activated sludge chamber and the sedimentation tank. It is expected that this will achieve a more complete denitrification than found in the fresh water circuits.

Using Wuhrmann's (1957) suggestion, a non-aerated denitrification sludge tank containing a slowly rotating motor-driven agitator was incorporated into the system next to the nitrification stage. In this tank, under more or less anaerobic conditions, heterotrophic denitrifying bacteria reduce nitrates to nitrogen gas which is given off (Barnard, 1973; Johnson and Schroepfer, 1964; Meske and Naegel, 1975; Mudrack, 1970; Mulbarger, 1971; and others). It may be necessary to provide an H^+ donor for the reduction of the nitrate, as during the active stage there is almost complete nitrification and almost all organic carbon compounds are either taken up by the bacterial cells or are oxydized to CO_2. Glucose and methanol are suitable for this purpose. In the purification of industrial sewage residues there are usually adequate H^+ donors in the influent.

Figure 27 shows an Ahrensburg salt water circuit which includes the additional denitrification stage. From the fish tank (1) water runs into a sump (2). Some of the water is pumped back untreated into the tank by a stainless steel or plastic centrifugal pump. Most of the water is pumped from the sump into the activated sludge tank (3). Here the sludge is aerated, often with pure oxygen to prevent the formation of excessive scum, and stirred. The sludge then flows into the funnel-shaped sedimentation tank (4), being introduced from the bottom. The construction of this tank is shown in detail in Figs. 28 and 29. The sludge settles in the internal cone-shaped reservoir from which it is sucked up by air pumps for return to the nitrification tank (3). The scum from the tank's surface is similarly air-pumped back to this tank. The clear, sludge-free, water partly flows into sump (2) and partly into the denitrification tank (5). The denitrification tank (5) is not aerated. It is gently stirred by a motor-powered agitator which is sited close to the tank's bottom. Tank M contains methanol — or glucose — which is dispensed in controlled doses into the denitrification tank, the methanol acting as H^+ donor. From the denitrification tank the sludge is returned to the activated tank (3).

The total volume of the circuit discussed here is about 2.5 m^3. It is operated with salt levels of between 20°/oo and 30°/oo. Water temperature was also varied experimentally with a mean of 25°C. The fish used in the experiments were tilapia species, i.e. *T. mossambica, T. aurea* and *T. nilotica*, and crosses between the last two species. The salt water circuits described above, and also circuits which had been slightly modified technically, have been operated experimentally for some years. Experience to date can be summarized as follows: It was possible to operate an activated sludge system with up to 35°/oo salinity for several months without significant problems arising. The ammonia values were found to be less than 1 mg/l, the nitrite values about 1 mg/l and the nitrate values about 300 mg/l. The growth rate of the fish in the experiments was good but not always as satisfactory as it might have been, mainly because of the aggressive behaviour of the fish.

Typical growth rates of tilapia in the salt water closed circuit tanks were as shown in Table 4. However it must be stressed that the operation of the activated sludge salt water purification system did not always perform as trouble-free as in the fresh water circuits. The particular problems which often arise are:

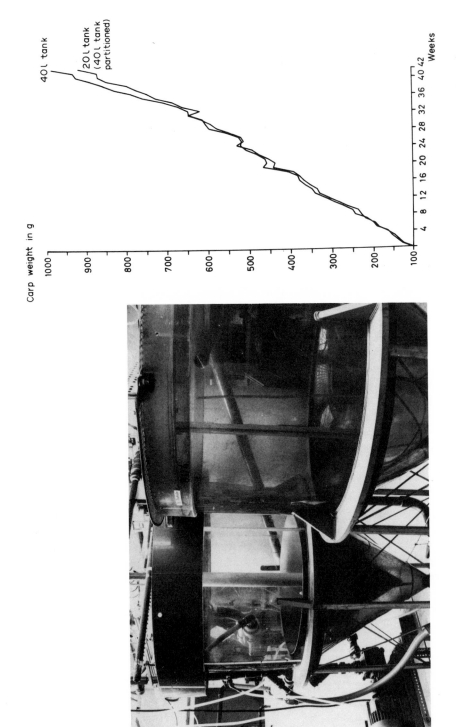

Fig. 29: External view of sedimentation tank as detailed in the section of Figure 28. The tank is made of transparent PVC and the internal cone-shaped reservoir and the central suction pipe can be clearly seen.

Fig. 30: Influence of tank volume (Experiment 67/20). Average growth rate of 20 carp, five each in 20 l tanks (lower curve) and five in a 40 l tank (upper curve) after 42 weeks.

TABLE 4

Growth rate of tilapia in salt water circuits

Date	Total weight (kg)	Mean individual fish weight (g)
Tank 1: *Tilapia aurea* X *Tilapia nilotica*		
17.03.78	1.00	3.00
24.05.78	9.44	37.70
28.06.78	14.98	84.40
Tank 1		
18.09.79	11.33	202.32
16.10.79	12.24	238.38
12.11.79	13.54	270.80
11.12.79	14.84	309.17
22.01.80	16.78	357.02
21.02.80	18.08	384.70
Tank 2: *Tilapia aurea*		
20.02.78	3.77	16.18
06.04.78	11.46	55.90
24.05.78	17.18	85.90
28.06.78	21.20	113.14

1. Because of the salinity there was a change in the relative density of the sludge, and in consequence a layer of sludge formed near the surface on occasions. This resulted in a dilution of the sludge concentration in the active tank, causing insufficient nitrification. This in turn may give undesirably high nitrite readings.

2. The nitrate value in the additional denitrification tank can be controlled so that the nitrate value remains at about 50 mg/l. Control was achieved by temporary disconnection of the tank from the circuit system and with the aid of hydrogen donors such as methanol. Nevertheless, it frequently performed biologically inefficiently on being re-connected to the circuit. Control of a continually working circuit presents difficulties when the denitrification tank is incorporated in the circuit all the time.

3. Until now the fish used in the experiments were tilapia (*T. aurea, T. nilotica* and their crosses). In spite of good feed uptake the growth rate of the fish in the high salinity environment was significantly lower when compared with tilapia raised in fresh water.

4. The need of the fish to adapt to salt water caused difficulties. As the tilapia in the experiments originate from fresh water aquaria there are problems associated with their adaptation to high salinity. Although the change was made over a period of 1 week there were recurring losses or alternatively progress was impaired.

The operation of a high-salinity activated sludge circuit depends on the performance of its micro-organisms. The activity of the bacteria is influenced by the abiotic environmental factors of temperature and salinity. As the water temperature rises the metabolism of the micro-organisms speeds up (e.g. Viehl, 1940). Correspondingly, biological open-air purification plants are more effective in summer than in the winter. As the Ahrensburg closed-circuit system, including its purification component, is completely enclosed and the water kept at a constant average over 20°C throughout the year, a very high rate of microbial activity can be expected.

The influence of salinity on microbial activity is not as well understood as that of temperature. According to Rheinheimer (1966) nitrification in brine proceeds at a less intensive rate than it does in fresh water. Reimann (1968) demonstrated that the nuclei of the oxidizing organisms of fresh water can adapt to the salinity of brine and continue to reduce organic substances. Ludzack and Noran (1965) reported that bacteria can, in general, adapt to various degrees of salinity, although there may be a prolonged period of adjustment before their reducing activity commences. This is also mentioned by Kincannon et al. (1966). Mudrack (1976) was able to confirm during laboratory experiments that activated sludge functions in conditions of salinity similar to that found in brine or brackish water. A prolonged period of adaptation and fairly constant salinity were, however, required.

The activated sludge units of the Ahrensburg experimental salt water circuits did, however, work more effectively when inoculated with North Sea mud. It is likely that the nitrification ability of marine micro-organisms took effect more or less immediately. Watson (1965) isolated and described *Nitrosocystis oceanus* as a marine nitrifying organism. The denitrification process in brine was reported by Thomas (1966) and other workers. Inoculated circuits display considerably fewer problems in their operation than found in those in which fresh water bacteria have to undergo the slow process of adaptation. The performance of the sludge is improved considerably (Meske and Rakelmann, 1981).

5. RESEARCH INTO THE MANAGEMENT, FEEDING AND BREEDING OF CARP IN WARM WATER

This chapter describes the experimental work on the carp (*Cyprinus carpio*) undertaken at Ahrensburg during recent years. The reports on the experiments are set out under these headings: Discussion, usually with graphs and statistics; Material and method; Summary of experimental data and results.

A large number of experiments was carried out. The experiments which are included here were chosen because they clearly illustrate a particular point. All of this chapter's experiments investigate the importance of the various factors influencing growth.

MATERIAL AND METHOD

START

On the day of the start of an experiment the fish were weighed and batched into groups.

END

End denotes the day of the last weighing of the fish and the cessation of the experiment.

MATERIAL (FISH)

Where a batch code number is given the experiment involved fish from pond cultures whose genetic background was unknown. Each number refers to a specific population. Where matings are referred to the fish were full siblings hatched at the Ahrensburg laboratory.

METHOD

The experiments were given code numbers indicating the year, consecutive experiment number in that year and the aquarium number, e.g.

Year	Experiment No.	Tank No.
67	72	1

FEED

The designation or description given — as for instance Trout concentrate or diet A — refers to a commercial bought-in concentrate. The formulation of this feed was specific to any given experiment. Hence in other experiments its analysis or composition may be quite different, although for reasons of simplicity it is described by the same designation.

A full analysis and the composition is only given for the experiments in feed quality.

FEEDING METHOD

Unless otherwise stated all fish in the experiments were fed daily — weekends and holidays included — between 07.30 and 16.30 hours.

Unless an alternative method is given feeding was in accordance with the scale of rationing shown in Table 5, at approximately 23°C water temperature.

TABLE 5

Mean weight of fish $\bar{x}$ (g)	Amount of feed as percentage of liveweight per day	Feeding frequency
— 10	10	10 times daily (hourly)
10 — 20	8	,,
20 — 30	6	,,
30 — 50	5	,,
50 — 100	4	,,
100 — 400	3.5	,,
400 — 600	3	,,
600 — 1000	2.5	5 times daily (2-hourly)
over 1000	2	,,

SYSTEM

Most of the experiments were completed in one of the Ahrensburg closed-cycle systems (p. 49), others in tanks with running temperated well water. The hydrobiological conditions were the same for all fish in an experiment. The experiments were with very few exceptions carried out in a greenhouse.

WATER FLOW RATE

The amount of water flowing through the aquaria and tanks per minute was controlled by a flow valve and checked at least once weekly.

WATER TEMPERATURE

Unless otherwise stated the experiments were carried out at an average water temperature of 23°C.

WEIGHING

Fish were weighed weekly unless otherwise stated. Generally all fish from a tank were weighed together, although in some cases — such as the last weighing at the end of an experiment — individual weights were established.

PRESENTATION OF EXPERIMENTAL DATA AND RESULTS

UP TO 1977

For the management and feeding experiments the weight and number of fish, average weight at the start and the end of an experiment are given for each treatment. From these values the total weight increase, the percentage weight increase, and the average weight increase are calculated. Where fatalities occurred during an experiment the increase and percentage increase of total weight of the losses were deducted from the weight at start of the experiment, i.e. total weight at start less the number of losses x average weight at start.

In the breeding experiments the weights of individual fish in each group were recorded at the start and at the end of experiments. From these data weight increases and percentage weight increases of each fish were established. Analysis of variance and t-test were used for statistical evaluation whenever possible.

AFTER 1977

The system was changed so that after 1977 the weights of fish up to the date of their death are included in the results.

MANAGEMENT

The experiments described here concern the influence on fish of:

> tank size (or volume);
> stocking rate (or density);
> water flow rate;
> light;
> temperature.

An overall summary of this sub-section can be found on p. 93.

INFLUENCE OF TANK VOLUME
EXPERIMENT 67/20

This experiment was conducted to check what effect tank size may have on the growth of fish. It involved 30 carp and ran for 42 weeks; 40 l plastic tanks were used, two being partitioned into two halves of 20 l each. Each of the 20 l halves had their own independent water supply and drainage.

Figure 30 illustrates that there is no disparity in the growth of the carp in the different-size tanks for the greater part of the experiment. A very small difference in the growth curve occurred after the thirtieth week when the length of the carp exceeded the dimension of the short side of the tanks at an individual weight of about 700 g. Even so the five carp in the 20 l tank continued to grow, and their weight at the end of the experiment was not statistically different from that of the carp in the 40 l tank. Figures 31 and 32 show the fish tanks at the end of the experiment.

The experiment proves that up to a certain point restriction of space* does not inhibit growth. The 20 l tank contained 4600 g of fish at the end of the experiment, which gives a fish:water ratio of 1:3. The rate of growth of the fish in the 20 l tank did however decline because of physical restrictions on their movements and difficulty in feeding.

MATERIAL AND METHOD

Duration	295 days
Fish	Carp of batch 12 66
Programme	67/20 — 1—4 : 5 carp each. Tank volume 20 l
	— 5 + 6 : 5 carp each. Tank volume 40 l
Feed	Trout concentrate A
Tanks	Plastic 20 l and 40 l
Water flow rate	2.5 l/min

---

*   See footnote on page 27.

Figs. 31 and 32: The effect of tank size (Experiment 67/20). *Top*: 40 l tank containing five carp, average weight 986 g, total weight 4930 g. *Bottom*: Two 20 l tanks, each containing five carp, average weight 919.2 g, total 9192 g. Both at end of 42-week experimental period.

DATA AND RESULTS

	67/20 − 1 to 4 Tank vol. 20 l	67/29 − 5 + 6 Tank vol. 40 l
At start		
Number	20	10
Weight (g)	2030	1015
$\bar{x}$ Weight (g)	101.5	101.5
At end		
Number	20	5
Weight (g)	18 385	4930
$\bar{x}$ Weight	919.2	986.0
Weight increase (g)	16 355	4425
Weight increase (%)	805.7	876.2
$\bar{x}$ Weight increase (g)	817.8	885.2
Feed consumption (g)	68 050	17 180
FCR*	4.16	3.88

Tank 67/20–5 suffered a technical fault and failed after 221 days.

*   Feed Conversion Ratio

THE INFLUENCE OF STOCK DENSITY
EXPERIMENT 68/54

This minor experiment demonstrates the strong dependency of carp fry growth on environmental factors.

100 carp fry which were put into a 30 l aquarium immediately after hatching grew nearly three times as fast as 1000 fry also in a 30 l aquarium. The fry in the densely stocked aquarium failed to develop as well as the carp fry in the less densely stocked one because the water flow rate was not proportionally adjusted. Both groups had the same *ad lib* feed regime (Fig. 33).

MATERIAL AND METHOD

Duration	21 days
Fish	carp (X 8056)
Programme	68/54—1 : 100 carp
	68/54—2 : 1000 carp
Feed	trout concentrate A and natural food
Tank	30 l glass aquaria
Water flow rate	2.5 l/min
Weighing	3-weekly

DATA AND RESULTS — SUMMARY

	68/53—1	68/54—2
Number at start	100	1000
Weight at start (g)		
$\bar{x}$ Weight at start (g)	newly hatched	
Number at end	77	877
Weight at end	155	682
$\bar{x}$ Weight at end (g)	2.01	0.78

THE INFLUENCE OF STOCK DENSITY
EXPERIMENT 67/73

This experiment with young carp confirmed that — tank size and water flow rate being the same for all groups — growth rate is inversely proportional to stocking rate.

With an average weight per fish of 3.9 g at the start of the experiment, 160 fish increased their weight by 441.4%, 80 by 640.1%, 40 by 757.1% and 20 by 906.5%. The fish in the tank with the lowest population density put on more than twice as much weight as those in the most densely stocked one (Fig. 34).

MATERIAL AND METHOD

Duration	78 days
Fish	carp (X 7025)
Programme	67/73—5 : 160 carp
	67/73—6 : 80 carp
	67/73—7 : 40 carp
	67/73—8 : 20 carp
Feed	trout concentrate A followed by concentrate B as from 30.01.68
Tanks	40 l, glass aquaria
Water flow rate	4 l/min

DATA AND RESULTS

	67/73—5 160 carp	67/73—6 80 carp	67/73—7 40 carp	67/73—8 20 carp
At start				
Number	160	80	40	20
Weight (g)	616	308	154	77
$\bar{x}$ Weight (g)	3.9	3.9	3.9	3.9
At end				
Number	160	79	40	20
Weight (g)	3335	2250	1320	775
$\bar{x}$ Weight (g)	20.8	28.5	33	38.8
Weight increase (g)	2219	1946	1166	698
Weight increase (%)	441.4	640.1	757.1	906.6
$\bar{x}$ Weight increase (g)	17	24.6	29.2	34.9
Feed consumption (g)	7920	4390	2255	1290
FCR	2.91	2.26	1.93	1.85

THE INFLUENCE OF WATER FLOW RATE AND STOCK DENSITY
EXPERIMENT 66/102

Experiment 66/102 involved tanks of the same size and volume, each containing four and eight carp respectively and with 2.5 l/min and 5 l/min water flow rates (WFR). The aim was to establish the relationship between water flow rate and stock density.

The graph (Fig. 35) plots the experiment's progress. It shows that eight carp with 5 l/min water rate have almost the same growth rate as four carp with 2.5 l/min. In contrast eight carp with a flow rate of only 2.5 l/min fell distinctly behind (see also Fig. 38).

MATERIAL AND METHOD

Duration	147 days	Feed	trout concentrate A
Fish	carp of batch 6 65		
Programme	66/102—1 + 2 : 4 carp each	Tank	40 l plastic
	WFR: 2.5 l/min	Water flow rate	see programme
	66/102—3 + 4 : 8 carp each		
	WFR: 2.5 l/min		
	60/102—5 + 6 : 8 carp each		
	WFR: 2.5 l/min		

DATA AND RESULTS

	66/102—1 + 2 WFR 2.5 l/min	66/102—3 + 4 WFR 2.5 l/min	66/102—5 + 6 WFR 5 l/min
At start			
Number	8	16	16
Weight (g)	1980	3950	3950
$\bar{x}$ Weight (g)	247.5	246.9	246.9
At end			
Number	8	16	16
Weight (g)	7120	11 120	13 495
$\bar{x}$ Weight (g)	890.0	695.0	843.4
Weight increase (g)	5140	7170	9545
Weight increase (%)	259.6	181.5	241.6
$\bar{x}$ Weight increase (g)	642.5	448.1	596.6
Feed consumption (g)	19 285	29 040	33 427
FCR	3.85	4.05	3.50

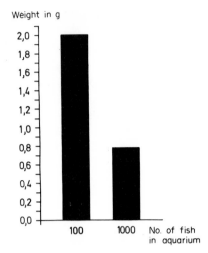

Fig. 33: The influence of stock density (Experiment 68/54). Average final weights of 100 and 1000 carp fry respectively after 3 weeks.

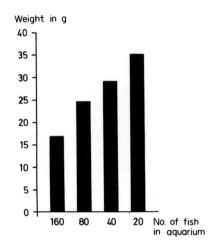

Fig. 34: The influence of stocking rate (Experiment 67/73). Average weight increases of young carp after 11 weeks in equal-volume tanks but with different stocking densities.

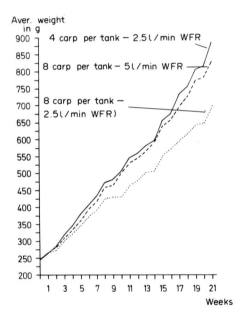

Fig. 35: The influence of water flow rate and stocking rate (Experiment 66/102). Average growth rate of carp at different stocking and water flow rates after 21 weeks. ———— Four carp in each tank, WFR 2.5 l/min; . . . . . . eight carp in each tank, WFR 2.5 l/min; – – – – – eight carp in each tank, WFR 5 l/min.

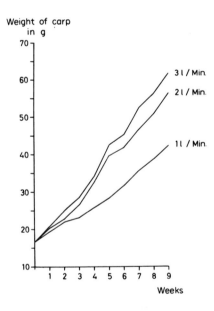

Fig. 36: The effect of water flow rate (WFR) (Experiment 68/81). Average growth rate of 50 carp at three different water flow rates over a 9-week period.

THE INFLUENCE OF WATER FLOW RATE (WFR) AND STOCKING RATE
EXPERIMENT 67/5

This experiment was similar in aim and method to experiment 66/102. In it, however, the fish weighed on average only 25.6 g at the start.

The experiment ran for 5½ months and the results differ slightly from 66/102 in that the weight differences between the groups with double the basic stock density and water flow were slightly greater than the corresponding values of 66/102. The order of the growth rates remains the same (cf Fig. 38).

MATERIAL AND METHOD

Duration	165 days
Fish	carp of batch 12 66
Programme	67/5—1 + 2 : 8 carp each
	WFR: 2.5 l/min
	67/5—3 + 4 : 16 carp each
	WFR: 2.5 l/min
	67/5—5 + 6 : 16 carp each
	WFR: 5 l/min
Feed	trout concentrate A
Tank	40 l plastic
Water flow rate	see programme

DATA AND RESULTS

	67/5—1 + 2 WFR 2.5 l/min	67/5—3 + 4 WFR 2.5 l/min	67/5—5 + 6 WFR 5 l/min
At start			
Number	16	32	32
Weight (g)	410	820	820
$\bar{x}$ Weight (g)	25.6	25.6	25.6
At end			
Number	16	32	32
Weight (g)	3370	5220	5665
$\bar{x}$ Weight (g)	210.6	163.1	177.0
Weight increase (g)	2960	4400	4845
Weight increase (%)	722.0	536.6	590.8
$\bar{x}$ Weight increase (g)	185.0	137.5	151.4
Feed consumption (g)	8585	11 170	11 555
FCR	2.90	2.54	2.38

THE INFLUENCE OF WATER FLOW RATE ON GROWTH
EXPERIMENT 68/81

This experiment showed that growth rate was positively correlated to the water flow rate. By the 9th week the experiment showed the strong influence of water flow rates at 1, 2 and 3 l/min respectively, all tanks being of the same size and volume and managed under the same feeding regime. The weight increase at 3 l/min was 268% and at 1 l/min only 152.2% (Fig. 36; cf. Fig. 37).

MATERIAL AND METHOD

Duration	64 days	Feed	trout concentrate A
Fish	carp mating 8045	Tank	40 l glass
Programme	68/81−1 : 50 carp	Water flow rate	see programme
	WFR: 3 l/min		
	68/81−2 : 50 carp		
	WFR: 2 l/min		
	68/81−3 : 50 carp		
	WFR: 1 l/min		

DATA AND RESULTS

	68/81−1 WFR 3 l/min	68/81−2 WFR 2 l/min	68/81−3 WFR 1 l/min
At start			
Number	50	50	50
Weight (g)	841	841	841
$\bar{x}$ Weight (g)	16.8	16.8	16.8
At end			
Number	50	50	50
Weight (g)	3095	2820	2125
$\bar{x}$ Weight (g)	61.9	56.4	42.5
Weight increase (g)	2254	1979	1284
Weight increase (%)	268.0	235.3	152.7
$\bar{x}$ Weight increase (g)	45.1	39.6	25.7
Feed consumption (g)	5940	5505	4930
FCR	2.64	2.78	3.84

THE INFLUENCE OF TANK VOLUME, WATER FLOW RATE AND STOCKING RATE
EXPERIMENT 67/72

Experiment 67/72 combined the previous separate experiments on the influence of tank volume, the water flow rate and stocking density. Seven tanks with a total of 340 fish were involved to test the effect of these parameters on the growth rate. The plan and the results of the experiment are visually represented in Fig. 37. The previous experiments are presented in the same manner in Fig. 38. The rectangles represent tanks. Their sequence corresponds to the growth rate of the fish within a specific experiment, starting with the highest weight increase. The relationship of the parameters on the growth rate of carp can be seen thus:

1. At basic stock density n, three times the unit water flow rate gives better growth than twice the unit water flow rate. Twice the unit water flow rate produces better growth than single unit flow rate only (Fig. 38).
2. At standard water flow rate growth is inversely proportional to increased stock density.
3. At unit standard water flow rate better growth is observed at stocking rate n than is seen when the stocking rate is a multiple of n (Fig. 38: experiments 68/54; 67/73; 66/102; 67/5; Fig. 37: experiment 67/72 − 8 to 14).
4. Growth rate is slightly better in large tanks than in smaller tanks. Experiments 67/20 (Fig. 37) and 67/72 (Fig. 37) would appear to confirm this. However, by examining the course of 67/20 it can be seen that growth in the smaller tank was only retarded when stock density became extreme and the length of the fish impeded their movement (Fig. 34).

Summing up the results of these experiments it is seen that the water flow rate and stock density are essential factors in the growth of carp. During experiments, as in commercial warm water fish production, maximum growth can be achieved if these two parameters are given sufficient attention and are properly balanced. The size of the tank is of minor importance.

MATERIAL AND METHOD

Duration	153 days
Fish	carp mating 7025
Programme	67/72−8 : 40 carp
	WFR: 2.5 l/min
	67/72−9 : 40 carp
	WFR: 5 l/min
	67/72−10 : 80 carp
	WFR: 5 l/min
	67/72−11 : 40 carp
	WFR: 5 l/min
	67/72−12 : 40 carp
	WFR: 2.5 l/min
	67/72−13 : 80 carp
	WFR: 2.5 l/min
	67/72−14 : 20 carp

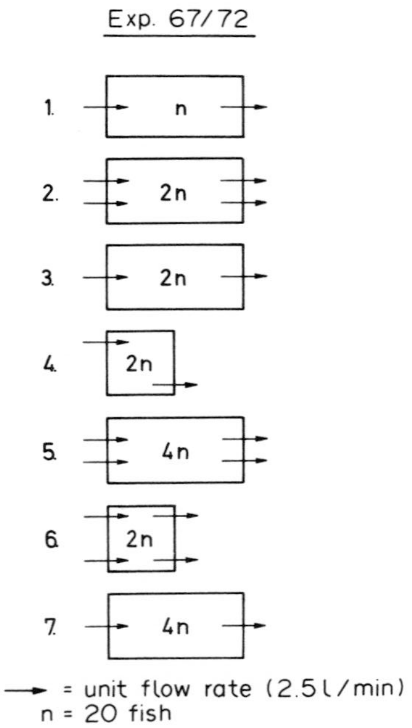

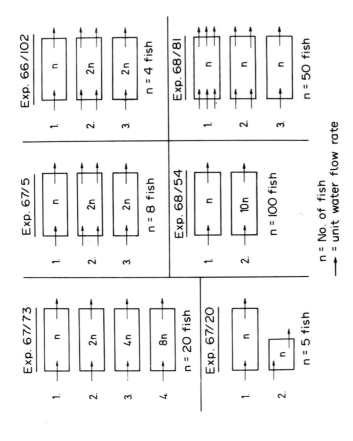

Fig. 38: The influence of tank volume, water flow rate and stocking rate. Diagrammatic presentation of results from experiments on the influence of the above three parameters, singly and combined. The rectangles represent tanks. They are ranked in order of growth achieved with best results at top. n is the number of fish per tank. Each arrow refers to a specified water flow rate as given in the description of the various experiments (see also Figure 37).

Fig. 37: The influence of tank volume, water flow rate and stocking rate (Experiment 67/72). The rectangles represent tanks and their sequence (top to bottom) ranks the growth rates achieved (highest first). n is the unit number of fish (20). One arrow represents unit flow rate (2.5 l/min). For details see text (cf. Figure 38).

Feed   WFR: 2.5 l/min
trout concentrate A; from 17.4.68 Trout concentrate B

Tank   Plastic
Water flow rate   As above

DATA AND RESULTS

	67/72–8 20 l tank WFR 2.5 l/min	67/72–9 20 l tank WFR 5 l/min	67/72–10 40 l tank WFR 5 l/min	67/72–11 40 l tank WFR 5 l/min	67/72–12 40 l tank WFR 2.5 l/min	67/72–13 40 l tank 2.5 l/min	67/72–14 40 l tank 2.5 l/min
At start							
Number	40	40	80	40	40	80	20
Weight (g)	150	150	300	150	150	300	75
$\bar{x}$ Weight (g)	3.8	3.8	3.8	3.8	3.8	3.8	3.8
At end							
Number	36	38	74	38	37	72	20
Weight (g)	1830	1685	3695	2495	2160	3015	1345
$\bar{x}$ Weight (g)	50.8	44.3	49.9	65.7	58.4	41.9	67.2
Weight increase (g)	1695	1543	3417	2353	2021	2745	1270
Weight increase (%)	1255.6	1086.6	1229.1	1657.0	1454.0	1016.7	1693.3
$\bar{x}$ Weight increase (g)	47.0	40.6	46.2	61.9	54.6	38.1	63.5
Feed consumption (g)	4085	3970	8210	5120	4655	7730	2780
FCR	2.41	2.57	2.40	2.18	2.30	2.82	2.19

THE EFFECT OF INTERRUPTED WATER FLOW
EXPERIMENT 68/41

This 22-week-long experiment shows the negative influence of interrupted water flow on 40 carp.

Although the flow was only cut off three times a week at night the fish under this regime displayed a statistically significant reduced growth rate compared to fish with continuous water flow (Fig. 39). Tanks 41–2 and 41–3 were supplied with water for a total of 2739 hours. Tanks 41–1 and 41–4 for 3696 hours. This gives a ratio of about 3:4, which compares with the average weight gains of the fish in g as 2:3.

This experiment demonstrates clearly the importance of a continuous water flow or renewal in warm water fish culture. Commercial fish farming enterprises frequently lack a continuous supply of warm water. The results obtained in the experiment indicate the need to ascertain the continuous availability of the right quantities of waste heat when setting up a warm water fish farm.

MATERIAL AND METHOD

Duration	154 days
Fish	carp from mating 7025
Programme	68/41–1 and 4: 10 carp each – continuous flow
	68/41–2 and 3: 10 carp each – water flow discontinued three times weekly – from 17.00 hours to 07.30 hours
Feed	trout concentrate A
Tank	40 l, plastic
Water flow rate	2.5 l/min

DATA AND RESULTS

	68/41–1 and 4 continuous water flow	68/41–2 and 3 water flow interrupted three times weekly
At start		
Number	20	20
Weight (g)	1660	1665
$\bar{x}$ Weight (g)	83.0	83.3
At end		
Number	19	12
Weight (g)	7360	3505
$\bar{x}$ Weight (g)	387.4	292.1
Weight increase (g)	5783	2506
Weight increase (%)	366.7	250.9
$\bar{x}$ Weight increase (g)	304.4	208.8
Feed consumption (g)	18 735	14 005
FCR	3.24	5.59

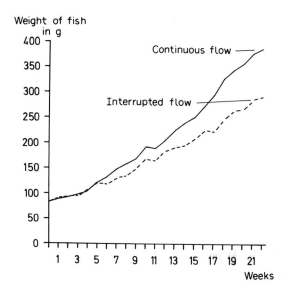

Fig. 39: The effect of interrupted water flow (Experiment 68/41). Average growth rate of 20 carp each with continuous (————) and interrupted (– – – – –) water flow (3 nights a week) over a period of 22 weeks.

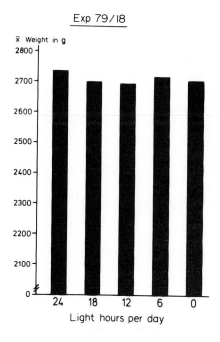

Fig. 40: The influence of light on carp weight (Experiment 79/18).

THE INFLUENCE OF LIGHT

Eight dark rooms measuring 2 x 2 x 2 m were constructed in one of the buildings of the Ahrensburg Institute to study the influence of light on the growth of table fish in greater detail. Each room contained two 500 l tanks which were connected to the warm water circuit. This provided the same water temperature and quality for all fish in the experiments. Each room was fitted with identical strip lighting and electrically operated feeders. The lights and feeders were wired to a central console with time switches controlling the lighting and feeding regimes.

Initially the influence of light on carp growth was investigated. The fish were exposed to light for 24, 18, 12, 6 and 0 hours daily respectively at 250 lux of white light. The feeders were scheduled to dispense feed for 6 hours each day, the feeding to coincide where appropriate with light periods.

Experiment 79/18 confirmed previous tests showing that light and darkness have no influence on the growth of carp. Figure 40 illustrates that after 4¾ months the growth of the carp was nearly the same irrespective of daylight length ranging from total darkness to all-day light.

THE INFLUENCE OF LIGHT ON CARP (Cyprinus carpio) GROWTH
EXPERIMENT 79/18

MATERIAL AND METHOD

Duration	139 days
Fish	carp (X 8022)
Programme	
Lighting	lighting commenced daily at 08.00 hours using two 40 W fluorescent tubes (L 40 W/39 warm white Osram Interna) per room for the following periods:

79/18—1 + 2—	24 hours of light per day
3 + 4	18 hours of light per day
5 + 6	12 hours of light per day
7 + 8	6 hours of light per day
9 + 10	0 hours of light per day

each tank contained 10 fish.

Feed and feed method	trout concentrate. Analysis: protein: 47%, oil: 8%, fibre: 3.5%, ash: 10%. fed to rationing scale given in Table 5 on p. 68. six times daily at hourly intervals starting at 08.30 hours by automatic feeder
Tank	0.4 m^3 (1 x 1 x 0.4 m depth) GRF
Water temperature	approximately 22.8°C
Water flow rate	6—7 l/min

DATA AND RESULTS

Experiment code No.	Daily lighting period (hours)	At start			At end			Weight increase		Feed consumed	
		No. of fish	Total weight (g)	$\bar{x}$ weight (g)	No. of fish	Total weight (g)	$\bar{x}$ weight (g)	(g)	(%)	Total (g)	$\bar{x}$/fish (g)
1 + 2	24	20	21 391	1069.6	20	54 736	2736.8	33 345	155.9	108 297	5414.9
3 + 4	18	20	21 571	1078.6	17	45 900	2700	24 329	112.8	88 557	(5209.2)
5 + 6	12	20	21 281	1064.1	19	51 175	2693.4	29 894	140.5	104 387	(5494.1)
7 + 8	6	20	21 550	1077.5	20	54 275	2713.8	32 725	151.9	106 918	5345.9
9 + 10	0	20	21 465	1073.4	20	54 025	2701.3	32 560	151.7	109 107	5455.4

THE INFLUENCE OF LIGHT ON GONAD DEVELOPMENT OF CARP
EXPERIMENT 80/5

This experiment was conducted to demonstrate experimentally the influence of light on gonad development. The fish were exposed to the same light intensity for varying periods of time. The result was unexpected. Short-day exposure of 6 and 12 hours daily had a clear and substantially beneficial effect on gonad growth. Long-day lighting of 18 and 24 hours duration a day inhibited gonad development (Fig. 41).

As in other experiments it was found that the growth of the fish was not influenced by the length of the light period.

MATERIAL AND METHOD

Duration	83 days
Fish	carp (X 9001)
Programme	80/5−1 + 2 24 hours of light per day
	3 + 4 18 hours of light per day
	5 + 6 12 hours of light per day
	7 + 8 6 hours of light per day
	9 + 10 0 hours of light per day
	each tank contained 10 fish
Lighting	lighting commenced at 08.00 hours each day with two 40 W light strips (L 40 W/39 warm white Osram Interna)
	brightness: about 110 lux on tank bottom and about 210 on water surface
Tank	0.2 m³ (1 × 1 m surface area, 0.2 m depth) GRF
Water temperature	23−24°C
Water flow rate	7 l/min approximately
Weighing	every 14 days
Feeding	trout feed. Analysis: protein: 47%, oil: 8%, fibre: 3.5% ash: 10%, fed according to Table 5 on p. 68, hourly, six times a day by auto-feeder;
	starting with 2.5% of liveweight as the mean, weight of fish in the experiment ranged from 600 to 1000 g.

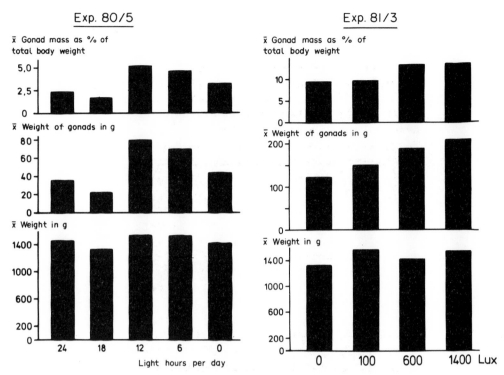

Fig. 41: The influence of light on gonad development (Experiment 80/5).

Fig. 42: The influence of light intensity on gonad development in carp (Experiment 81/3).

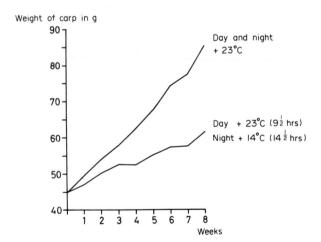

Fig. 43: The influence of water temperature fluctuations (Experiment 68/23). Average growth rate of 20 carp with constant water temperature at 23°C (upper curve) and 20 carp with night water temperature reduced to 14°C (lower curve). Duration of experiment: 8 weeks.

DATA AND RESULTS

Experiment Code No.	Daily light period (hours)	Fish at Start			Fish at end			Weight increase		Feed fed	FCR	Fish			Gonad weight		Number		Not determined
		No.	Total weight (g)	Mean weight x̄ (g)	No.	Total weight (g)	Mean weight x̄ (g)	(g)	(%)	(g)		Length (cm)	Height (cm)	Total (g)	Mean x̄ (g)	As mean percentage of LW x̄ (%)	♀	♂	
1 + 2	24	20	14 217	710.85	19	27 736	1459.8	14 939	105.08	34 115.7	2.30	41.30	14.3	685.0	35.7	2.44	9	9	2
3 + 4	18	20	13 621	681.05	18	23 878	1326.6	12 054	88.50	31 249.7	2.59	40.55	13.7	386.9	21.9	1.67	7	11	2
5 + 6	12	20	14 041	702.05	19	28 934	1522.8	15 955	113.63	33 745.8	2.12	41.70	14.3	1522.8	79.6	5.23	7	12	1
7 + 8	6	20	14 473	723.66	18	27 476	1526.4	14 803	102.28	33 239.1	2.25	41.80	13.8	1238.8	69.7	4.56	8	9	3
9 + 10	0	20	14 096	704.82	17	24 047	1414.5	14 053	99.69	32 576.7	2.32	40.60	13.9	763.8	44.4	3.17	6	11	3

THE INFLUENCE OF LIGHT INTENSITY ON GONAD DEVELOPMENT IN CARP
EXPERIMENT 81/3

The previous experiment (80/5) examined the influence of light period length on gonad development in carp. This experiment was carried out to ascertain the effect of light intensity on gonad development. Total darkness and three light intensity levels of 100, 600 and 1400 lux were compared. The gonad weight increase was found to be in relation to higher light intensity (Fig. 42). As in other experiments light had no influence on feed consumption or growth rate.

   This would appear to be of interest to commercial carp producers. However in the case of brood carp, light – in terms of intensity and duration – plays an essential role in gonad development, higher gonad weight increases occuring at higher light intensities.

MATERIAL AND METHOD

Duration   97 days

Fish   carp (X 9010)

Programme

	Lighting period	light intensity (lux)
81/3—1 + 2	0 hours per day	0
3 + 4	12 hours per day	100 lux striplight
		L 40 W/39 Interna
		warm white Osram
		110 cm above water surface
5 + 6	12 hours per day	600 lux approx. Two 400 W mercury
		vapour lamps
		95 cm above water surface
7 + 8	12 hours per day	1400 lux approx. Two 400 W mercury
		vapour lamps
		55 cm above water surface

                  5 fish per tank
                  lighting commenced at 08.30 hours each day

Tanks   200 l (1 × 1 × 0.2 m depth) GRF

Water temperature 24.2 ± 1.2°C

Water flow rate   10 l/min approximately

Weighing   every 14 days

Feed and   trout feed. Analysis: protein: 47%, oil: 8%, fibre: 3.5%, ash: 10%,
feed method   12 hourly feed per day by automatic feeder, rationed as per Ahrensburg
                  feed scale (page 68).

DATA AND RESULTS

Experiment code No.	Light intensity (lux)	Fish numbers and weights						Weight increase		Feed fed (g)	FCR
		at start			at end						
		No.	Total (g)	Mean (g)	No.	Total (g)	Mean (g)	(g)	(%)		
1 + 2	0	10	5964	596.4	10	13 131	1313.1	7167	120.17	18 527	2.59
3 + 4	100	10	5978	596.4	10	15 577	1557.7	9599	160.57	20 163	2.10
5 + 6	~ 600–700	10	5978	597.8	10	14 231	1423.1	8253	138.06	19 792	2.40
7 + 8	~1200–1500	10	5989	598.9	10	15 375	1537.5	9386	156.72	20 033	2.13

Experiment code No.	Light intensity (lux)	Fish size		Gonad weights			No. of fish by sex		Gonad weights by sex							
		Length (cm)	Depth (cm)	Total (g)	Mean (g)	As mean percentage of liveweight	♂	♀	Total (g)		Total (%)		Mean (g)		As mean percentage of liveweight	
									♀	♂	♀	♂	♀	♂	♀	♂
1 + 2	0	41.3	13.4	1234	123.4	9.40	5	5	813.0	421.0	65.9	34.1	162.6	84.2	12.4	6.4
3 + 4	100	42.8	14.8	1500	150.0	9.63	7	3	449.0	1051.0	29.9	70.1	149.7	150.1	9.3	9.8
5 + 6	600–700	41.7	13.8	1891	189.1	13.29	3	7	1450.0	441.0	76.7	23.3	207.1	147.0	14.2	10.9
7 + 8	1200–1500	43.1	14.3	2093	209.3	13.61	5	5	1219.0	874.0	58.2	41.8	243.8	174.8	15.7	11.5

THE INFLUENCE OF WATER TEMPERATURE VARIATION ON CARP
EXPERIMENT 68/23

This experiment was to examine — among other things — whether the warm water culture of carp is possible with industrial cooling water whose temperature is not constant. A constant 23°C is a proven, favourable temperature and this was compared in an experiment with regularly alternating water temperatures of 23°C and 14°C. This corresponds to cooling water temperature of many processes and factories with reduced activity at night, i.e. power stations, dairies, distilleries, etc.

The statistically proven difference in the growth rate of fish kept at a constant water temperature of 23°C and at changing temperatures of 23°C and 14°C is convincing: alternating water temperature has a markedly deleterious effect on carp growth (Fig. 43). Although all groups were fed only during the day when all fish were kept at 23°C, those suffering abrupt temperature drop of 9°C every evening could only show a weight increase of 36.7%. The fish kept at a constant 23°C in contrast gained 89.4%.

Even allowing for a possible energy cost saving the reduced performance of the fish crop through intermittent warm water supply represent a considerable loss in the potential profit for a commercial enterprise.

MATERIAL AND METHOD

Duration	57 days
Fish	carp from mating 7025
Programme	68/23—1 and 2: 10 carp each
	water temperature at 23°C for 24 hours
	68/23—5 and 6: 10 carp each
	water temperature at 23°C, 7.30—17.00 hours
	and at 14°C, 17.00 to 7.30 hours
Feed	trout concentrate A
Tank	40 l, plastic
Water flow rate	2.5 l/min

DATA AND RESULTS

	68/23−1 and 2 24 hours at 23°C water temperature	68/23−5 and 6 9½ hours at 23°C, 14½ hours at 14°C water temperature
At start		
Number	20	20
Weight (g)	900	900
$\bar{x}$ Weight (g)	45	45
At end		
Number	20	20
Weight (g)	1705	1230
$\bar{x}$ Weight (g)	85.2	61.5
Weight increase (g)	805	330
Weight increase (%)	89.4	36.7
$\bar{x}$ Weight increase (g)	40.2	16.5
Feed consumption (g)	2495	2540
FCR	3.10	7.70

MANAGEMENT: SUMMARY OF EXPERIMENTS

In this series of experiments on management several environmental factors — tank size and volume, stock density, water flow rate, light and water temperature — were examined in their relation to the growth of carp. Certain aspects of the work described here require further research but some conclusions can be drawn from the data: carp react to the quality — in the widest sense — of the water available to them.

The water exchange rate in the tanks must therefore be adjusted in relation to stock density. Only continual water change ensures a good growth rate. When the water flow is properly adjusted the tank size or its volume is of minor importance.

The influence of light and gonad maturation has been demonstrated. The growth of carp is not influenced by light.

Fluctuations in the water temperature proved distinctly disadvantageous in relation to the growth of carp.

FEEDING

This section describes experiments in aquaria concerned with feed quality, feed quantity and feeding method.

RAISING CARP FRY USING VARIOUS FEEDS
EXPERIMENT 69/24

A considerable number of experiments have shown that nauplii of the Californian brine shrimp, *Artemia salina*, are the most satisfactory feed for newly-hatched carp fry in aquaria. Experiment 69/24 was conducted to ascertain whether other feeds can be used for fry.

200 fry in each of the tanks 24 — 1 to 24 — 5 were fed respectively:

A: dried algae powder (*Scenedesmus* spp.)

B: brine shrimp (*Artemia salina*)

C & D: two bought-in trout feeds

E: skimmed milk curd

On the 17th day and 28th day, the end of the experiment, the fish in all the tanks were counted and weighed in a vessel of known water content. The results of this 4-week-long experiment were unequivocal (Fig. 44): only live feed B gave both a satisfactory growth rate ($\bar{x}$ weight 972 mg) and a good survival rate (71%). All other feeds resulted in high or total mortality.

At the same time 200 carp fry in tanks 24 – 6 to 24 – 10 were fed *Artemia salina* for the first 2 weeks of the experiment before being switched to feeds A, B, D and E as set out above. The fish of one tank (24 – 7) were changed over from *Artemia* to live *Chironomus* larvae (F) after 4 weeks. Figure 45 shows the average weights and the losses after 4 weeks. This demonstrates clearly that live feed (B/F) alone is the most suitable diet for carp fry when compared with the alternatives tested. Feeding algae (A) and skimmed milk (E) after *Artemia* reduced the magnitude of the losses (21.5% and 34% respectively) but also gave relatively smaller weight increases. The highest losses were experienced with the bought-in trout feeds C and D following *Artemia* at 51.5% and 80% respectively.

This experiment confirmed that carp fry raised in aquaria are best fed live food. After hatching the fry is hardly capable of taking up food floating on the surface or lying at the bottom. Only food actually floating in front of the mouth of the fish will be eaten. Experience at Ahrensburg suggests this condition is best met by *Artemia* larvae. After 2 weeks the food can be changed, although calculations must allow for any losses.

During recent years a number of experiments have been carried out to find a substitute for *Artemia salina* larvae as a starter for carp fry. Several commercial feeds and experimental formulations, fed as fine meal or small granules, as used for salmonids and aquaria fish, were investigated. These experiments are not described here because in principle they all gave the same results as Experiment 69/24. No viable substitute for live *Artemia* has so far been found. Even when the fry took up the offered dry feed no satisfactory results were achieved. The fish grew poorly, very frequently displayed deformities, and for the most part died. On the other

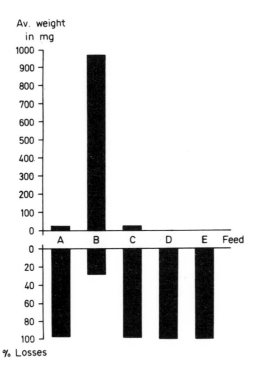

Fig. 44: Raising carp dry using various feeds (Experiment 69/24–1 to 5). Average weights and losses of carp fry at 4 weeks. 200 carp fry per group. A: dried algae powder (*Scenedesmus spp.*); B: brine shrimp (*Artemia salina*); C + D: two bought-in trout concentrates; E: skimmed milk curd.

Fig. 45: Rearing newly hatched carp fry using different feeds (Experiment 69/24–6 to 10). Average weight and percentage losses of 200 carp in each group at 4 weeks. All groups were fed brine shrimp (*A. salina*) (B) exclusively for 2 weeks followed for 2 weeks by B/A: powered algae (*Scenedesmus spp.*); B/F: insect larvae (*Chironomus*); B/C: trout concentrate A; B/D: trout concentrate B; B/E: skimmed milk curd.

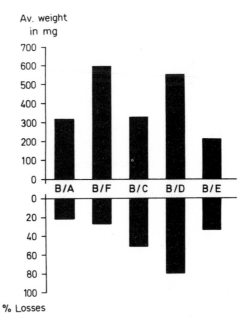

hand their sibs fed *Artemia* exhibited good and constant growth, almost no deformities and only small losses.

Von Lukovicz and Rutkowski (1976) were also unsuccessful in raising carp fry optimally on dry feed. Feeding *Artemia* at least for the first few days gave the best results. Similarly Anwand *et al.* (1976) report that live natural feed must be fed for at least a few days before the fish can be switched to a dry feed. They experimented with several formulations. The result was high losses and poor growth.

MATERIAL AND METHOD

Duration	27 days
Fish	carp from mating 9001
Method	69/24—1: 200 carp fed powdered algae (*Scenedesmus* spp.)
	69/24—2: 200 carp fed *Artemia salina*, the brine shrimp
	69/24—3: 200 carp fed trout concentrate A
	69/24—4: 200 carp fed trout concentrate B
	69/24—5: 200 carp fed skimmed milk curd*
	69/24—6 to 10 fed *Artemia salina* for 17 days and then given:
	69/24—6: 200 carp fed powdered algae
	69/24—7: 200 carp fed *Chironomus* larvae
	69/24—8: 200 carp fed trout feed A
	69/24—9: 200 carp fed trout feed B
	69/24—10: 200 carp fed skimmed milk curd*
Feed	Algae:
	Scenedesmus spp.
	Artemia salina:
	newly hatched larvae of the brine shrimp
	trout concentrate A:
	finely ground: crude protein. 38.0%; oil: 4.5%; fibre: 7.0%
	trout concentrate B:
	finely ground: crude protein. 46.6%; oil: 5.4%; fibre: 5.0%
	skimmed milk curd:
	in deep-frozen portions
Feed method	*ad lib*
Tank	10 l,Plexiglas
Water flow rate	1 l/min

---

* Cottage cheese. Analysis not given by author.

DATA AND RESULTS

Experiment code No.	69/24—1	69/24—2	69/24—3	69/24—4	69/24—5	69/24—6	69/24—7	69/24—8	69/24—9	69/24—10
Feed	Algae *Scenedesmus*	*Artemia salina*	Trout conc.	Trout conc.	Skimmed milk curd	Algae *Scenedesmus*	Midge larvae *Chironomus*	Trout conc.	Trout conc.	Skimmed milk curd
Feeding regime	A	B	A / C	B / D	E	A	F	B / C	B / D	E
At start										
Number	200	200	200	200	200	200	200	200	200	200
On 17th day										
Number	24	165	65	6	76	179	172	137	181	177
Weight (g)	0.85	22.30	0.90	0.01	0.80	27.30	21.80	15.50	23.00	21.40
$\bar{x}$ Weight (mg)	35	135	65	2	10	152	127	113	127	121
At end										
Number	4	142	2	0	0	157	145	97	40	132
Weight (g)	0.10	138.10	0.05	0	0	50.60	87.60	32.40	22.40	28.70
$\bar{x}$ Weight (mg)	25	972	25	0	0	322	604	334	560	217

RAISING YOUNG CARP WITH DAIRY PRODUCT BASED FEEDS AND WITHOUT FISH MEAL
EXPERIMENT 72/22

Experiment 72/22 was designed to develop a carp diet without fish meal. Its purpose was also to test feed components readily available in sufficient quantities. Whey was chosen because this by-product of cheese-making is frequently available as a liquid in considerable quantities which cannot be economically utilized. It also presents disposal problems. Feed mixes SM1 to SM4 were formulated on the basis that the amino acids of a feed should be matched to the amino acid composition of the fish being fed as closely as possible. This was thought particularly important in experiments with young fish.

Hence 8-week-old carp were analysed for their amino acid composition (method: Meske et al., 1978). Soya was used as a protein source apart from powdered dried whey, and therefore the amino acid contents of both were determined. The difference between amino acids in the feed and the fish was ascertained, the results showing that a feed providing 74% of the protein from soya and 26% of the protein from dried whey would meet the young carp's needs. The analyses demonstrated a marked deficiency in methionine and lysine and the feed mixes were augmented accordingly.

Following this work feed mixes SM1 to SM4 were formulated. Their composition is given under feed formulations on page 99.

As an added source of energy 10% fat in the form of anhydrous soya oil extract — melting point $43°C$ — was incorporated in the feed. All mixtures were supplemented with a vitamin premix at 0.5%. Feeds SM3 and SM4 also contained additional trace elements. Data on both the vitamin and trace element admixes are described in Meske et al., 1977. Feeds SM2 and SM4 were supplemented with methionine and lysine.

Mixes SM1 to SM3 were fed to young carp each weighing 0.27 g at the start. The experimental evidence showed that SM3 achieved the best results (Fig. 45) and that SM4 (not shown) showed no growth improvement on SM3. SM4 included a massive lysine supplementation and taking cost into consideration the use of this mix was discontinued.

Feed mix E, containing fish meal, gave the highest individual weight gains but this was offset by the significantly higher incidence of fatalities.

The second part of experiment 72/22—9 to —11 used carp with individual initial weights of 8 g fed the same feed formulations as before. The results are illustrated in Fig. 47. They confirm SM3 as the best feed producing even better growth than the fish meal containing control feed E. This superiority can also be seen in Figs. 48 and 49, which show the spread of the individual weights of the fish on SM3 and E. The SM3 population was more uniform in size, an important commercial consideration when variation is undesirable.

The fish meal-free SM3 formulation was also examined for toxic substances. This showed that SM3 contained significantly lower levels of mercury and lead than the three commercially available concentrates tested. In contrast with these concentrates too, no traces of aflotoxins, γ, γ DDT, DDE, DDD and HCB were found in SM3.

The experiment demonstrated that it is possible to rear young carp without fish meal. In the following experiments this was examined further by investigating other protein sources which could replace expensive fish meal which is often in short supply.

MATERIAL AND METHOD

Duration	62 days
Fish	carp of cross 2047 (72/22—1 to 8)
	carp of cross 2040 (72/22—9 to 12)

Method		
	72/22—1 and 2:	100 carp in each aquarium fed SM1
	72/22—3 and 4:	100 carp in each aquarium fed SM2
	72/22—5 and 6:	100 carp in each aquarium fed SM3
	72/22—7 and 8:	100 carp in each aquarium fed E
	72/22— 9:	10 carp fed SM1
	72/22—10:	10 carp fed SM2
	72/22—11:	10 carp fed SM3
	72/22—12:	10 carp fed E

Feed formulation		SM1	SM2	SM3	SM4
	Soya (g)	391	391	391	391
	Whey powder (g)	509	509	509	509
	Fat	100	100	100	100
	Vitamin pre-mix (g)	5	5	5	5
	D,L-Methionine (g)	—	1.43	1.43	1.43
	L-Lysine (g)	—	4.79	4.79	23.8
	Trace Elements (g)	—	—	0.75	0.75

Analysis					
	Crude protein (%)	23.5	24.4	24.4	26.3
	Oil (%)	10	10	10	9.8
	Fibre (%)	2.5	2.5	2.5	2.5

Feed E

25%	Fish meal
26.4%	Barley
10%	Wheat
7%	Oats
5%	Maize (corn)
10%	Bone meal
5%	Shrimp meal
5%	Dried milk (spray-dried)
2.5%	Yeast
2%	Calcium
1%	Salt
0.5%	Vitamin pre-mix
0.2%	Colouring matter
0.3%	Binder

100%

Analysis

Crude protein	36.0%
Oil	5.0%
Fibre	3.5%

SM1, SM2 and SM3 were fed in powder form. Feed E is a commercially available trout concentrate.

Feed frequency	Hourly by hand from 7.30 hours to 17.00 hours.
Feed rationing and weighting	72/22—1 to 8: day 1 to 8 at 30% of liveweight per day
	day 9 to 19 at 20% of liveweight per day
	day 20 onwards 10% of liveweight per day
	72/22—9 to 12: up to 10 g $\bar{x}$ weight at 10% of liveweight per day
	10 g to 20 g $\bar{x}$ weight at 8% of liveweight per day
	20 g to 30 g $\bar{x}$ weight at 6% of liveweight per day
	30 g to 50 g $\bar{x}$ weight at 5% of liveweight per day
	50 g to 100 g $\bar{x}$ weight at 4% of liveweight per day
	The fish were weighed once a week and their ration adjusted accordingly
Tank	72/22—1 to 8: 20 l glass aquaria
	72/22—9 to 12: 40 l glass aquaria
Water and flow rate	spring, 1 l/min
Water temperature	between 22.2 and 25.2°C

DATA AND RESULTS

Experiment code No.	72/22–1 + 2	72/22–3 + 4	72/22–5 + 6	72/22–7 + 8	72/22–9	72/22–10	72/22–11	72/22–12
Feed	SM1	SM2	SM3	E	SM1	SM2	SM3	E
At start								
Number	200	200	200	200	10	10	10	10
Weight (g)	53.0	53.1	51.5	53.4	79.8	79.4	79.6	79.1
$\bar{x}$ Weight (g)	0.27	0.27	0.26	0.27	7.98	7.94	7.96	7.91
At end (62 days)								
Number	146	151	141	104	10	10	10	10
Σ Weight (g)	334.4	314.0	495.9	417.8	343.9	396.2	535.1	505.8
$\bar{x}$ Weight (g)	2.29	2.08	3.52	4.02	34.39	39.62	53.51	50.58
Σ Weight increase (g)	295.7	273.8	459.5	390.0	264.1	316.8	455.5	426.7
Weight increase (%)	764.1	681.1	1262.4	1402.9	331.0	339.0	572.2	539.4
Food conversion ratio	3.75	4.12	3.11	2.37	3.84	3.34	2.66	2.55

THE INFLUENCE OF ALGAL-RICH FEED ON CARP GROWTH
EXPERIMENT 68/14

Commercial production of unicellular algae could be of considerable importance in fish feeding. Unicellular algae of the genus *Scenedesmus* distinguish themselves by having a crude protein content of 55%. The algae also contain 13% carbohydrates, 12% oil, 12% ash and 8% fibre, as percentage of dry matter They are also rich in vitamin B_1 (12 γ/g), B_2 (40 γ/g) B_{12} (0.1 γ/g) and C(0.3–0.6 mg/g of dry matter).

Experiment 68/14 was conducted to test the effect of feeding these green algae to carp. It lasted 10 weeks and involved 160 fish. Two tanks of 20 carp each were given the following feed regimes:

pure algae powder	(tanks 14—4 + 8)
a mixture of 80% algae and 20% sugar	(tanks 14—2 + 6)
a mixture of 32% algae and 68% trout concentrate	(tanks 14—3 + 7)
dry trout concentrate only	(tanks 14—1 + 5).

The results of the experiment show that the 32% algae and 68% trout concentrate mixture achieved by far the best growth. It proved highly significant statistically compared to the three other feeds. The mixture also gave the best feed conversion ratio within the experiment of 1.93. The pure algae feed gave a growth rate which was significantly better than that of the trout concentrate regime.

The statistically significant results justify the statement that the feeding of *Scenedesmus* is of great practical value to carp.

It is possible that the high vitamin content of this alga is responsible for the positive results of these algae feed experiments. When formulating a complete and specific carp feed this material should be given full consideration.

MATERIAL AND METHOD

Duration	64 days
Subject	carp of mating 7025
Fish	68/14—1 and 5: 20 carp per group, trout concentrate A (control)
	68/14—2 and 6: 20 carp per group, 80% algae and 20% sugar
	68/14—3 and 7: 20 carp per group, 32% algae and 68% trout diet A
	68/14—4 and 8: 20 carp per group, algae, powdered algae (*Scenedesmus* spp.) reconstituted in water and sometimes with specified admixtures, put through mincer and dried
	Feed analysis: Trout diet A: 40.1% CP, 4.3% oil, 7.6% fibre
Tank	40 l, plastic
Water flow rate	2.5 l/min

DATA AND RESULTS

	68/14—1 and 5 Trout Diet A (control)	68/14—2 and 6 80% algae and 20% sugar mix	68/14—3 and 7 32% algae and 68% Trout Diet A mix	68/14—4 and 9 algae
At start				
Number	40	40	40	40
Weight (g)	1330	1330	1330	1330
$\bar{x}$ Weight (g)	33.3	33.3	33.3	33.3
At end				
Number	40	40	39	40
Weight (g)	2790	3275	4530	3410
$\bar{x}$ Weight (g)	69.8	81.9	116.2	85.3
Weight increase (g)	1460	1945	3233	2080
Weight increase (%)	109.8	146.2	249.3	156.4
$\bar{x}$ Weight increase (g)	36.5	48.6	82.9	52.0
Feed consumption (g)	4240	6535	6255	6100
FCR	2.90	3.36	1.93	2.93

TESTING FOR OPTIMUM ALGAE CONTENT IN CARP FEED
EXPERIMENT 74/8

An increasing percentage — from 0% to 90% — of *Scenedesmus* algae meal was admixed to the basic fishmeal-free SM3 formulation (see Experiment 72/22 page 99). The response from day 84 onwards showed similar tendencies for growth rate as found during the work on the grass carp, which is discussed in Chapter 6.

Figure 50 shows that as the algae content increases there is initially a favourable response for growth rate. At 50% the effect lessens and from 70% upwards heavy losses occur which at a 90% algae content reaches a mortality rate of 87%.

After the 84th day, the experiment — which had commenced with 1000 fish — was carried on only with those groups which received up to and including 60% algae. These groups continued to display the same growth characteristics as shown up to day 84.

The cause for the depression of growth and the fatalities in carp feeds with high algae content may be due to these mixtures being nutritionally unbalanced. If so it is a deficiency problem. Certainly the fish in this case did not suffer from heavy metal toxicity which may occur where algae are produced in the vicinity of industrial complexes. Analysis of their meat showed that carp kept for 1 year on the 68% SM3 – and 32% algae mix contained 0.21 ppm of lead and 0.007 ppm cadmium. The tolerance limits in fish are 0.5 ppm for lead

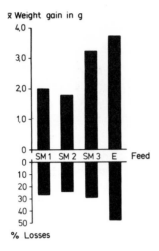

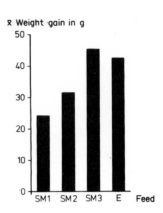

Fig. 46: The rearing of carp fry without fish meal (Experiment 72/22–1 to 8) (mixes SM1 to SM3–E as control). Weight increases are shown in upper columns; mortalities are given in lower columns; 11 at end of experiment (62 days). (Meske and Pfeffer, 1977). For mix composition see p. 99.

Fig. 47: The rearing of carp fry without fish meal (Experiment 72/22–9 to 12) (SM1 to SM3-E as control feed). Average weight gained at end of experiment (62 days). (Meske and Pfeffer, 1977). For mix composition see p. 99.

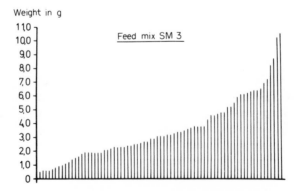

Fig. 48: Individual weight distribution in carp fry on fish meal-free feed mixture SM3 (Experiment 72/22).

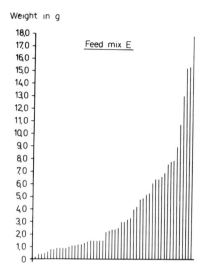

Fig. 49: Distribution of weights of carp fry fed control feed mixture E containing fish meal. (Experiment 72/22–8).

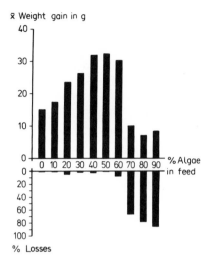

Fig. 50: Feeding algae (*Scenedesmus*) to young carp at 0% to 90% algae inclusion rates (Experiment 74/8). Upper columns: average weight increase (g); Lower columns: losses (%) (Meske and Pfeffer, 1977).

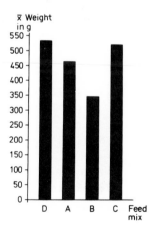

Fig. 51: Average weight of carp after 180 days on feed containing algae but no fish meal (D) and three fish meal-containing trout feeds (A, B, C) (Experiment 74/10) (Meske and Pfeffer, 1977).

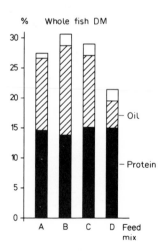

Fig. 52: Dry matter content of carp (whole) on 4 feed mixes (Experiment 74/10). D contains no fishmeal; A, B and C contain fish meal; hatched columns: oil; solid columns: protein (Meske and Pfeffer, 1977).

and 0.075 ppm for cadmium. Even the carp fed 80% algae was below these tolerance levels at 0.34 ppm lead and 0.063 ppm cadmium.

The protein efficiency ratio (PER) defines the weight increase in g achieved for each g of crude protein input. The highest PER was obtained in an experiment where the fish were fed a mixture of SM3 and algae. The results of this experiment are given in Table 6. The effect of the addition of yeast or milk albumen on growth can also be seen, but the significantly better PER was obtained with the feed containing algae compared to the formulation without algae.

TABLE 6

The relation between live weight gain and protein source

Feed mixture	PER
Trout concentrate X	1.10
SM3	1.08
68% SM3 + 32% algae	1.54
68% SM3 + 32% yeast	1.32
68% SM3 + 32% casein	1.25

Following the extremely successful results achieved with the inclusion of algae at rates up to 50%, experiment 74/10 was carried out with a mix of 68% fish meal-free basic SM3 and 32% dried green algae tested against three commercially available concentrates (containing fish meal) on carp.

MATERIAL AND METHOD

Duration	182 days			
Fish	carp of mating 40 02			
	74/8—1 + 2	50 carp each 100% SM3*	74/8—11 + 12	50 carp each 50% SM3 50% algae
	74/8—3 + 4	50 carp each 90% SM3 10% algae	74/8—13 + 14	50 carp each 40% SM3 60% algae
	74/8—5 + 6	50 carp each 80% SM3 20% algae	74/8—15 + 16	50 carp each 30% SM3 70% algae
	74/8—7 + 8	50 carp each 70% SM3 30% algae	74/8—17 + 18	50 carp each 20% SM3 80% algae
	74/8—9 + 10	50 carp each 60% SM3 40% algae	74/8—19 + 20	50 carp each 10% SM3 90% algae

Feed preparation	both the SM3 and algae powder mixes were moistened with water, put through a mincer and dried at 30°C. The feed was then ground till the particles were of suitable size. The dust in the feed was sieved off.
Feed regime	*ad lib* for first 28 days. Then as per ration table (p. 68) fed hourly (10 times per day)
Tanks	40 l, glass aquaria
Water flow rate	2.5 l/min approximately
Water temperature	24°C
Remarks	for the purpose of analytical work and in view of the frequent high incidence of loss in carp, particularly affecting those fed on mixes with high algae content, the following changes in the carp populations were made:

on the 84th day of the experiment 5 fish were removed from each
tank at random
on the 141st day of the experiment the 10 lightest fish were removed
from each tank
on the 155th day the experiment was continued with the 10 heaviest
fish from tanks 1 to 14

* For analysis of SM3 see p. 99.

SUMMARY OF DATA + RESULTS — Stage 1 (84 days)

Experiment code No.	Algae inclusion rate (%)	At start Number of fish	At start $\bar{x}$ Weight (g)	On day 84 Number of fish	On day 84 $\bar{x}$ Weight (g)	Weight gained ($\bar{x}$ g)	Weight gained (%)	FCR	Losses (%)
74/8— 1 + 2	0	100	1.81	99	17.08	15.27	842.5	3.46	1
3 + 4	10	100	1.83	99	19.33	17.50	956.4	3.28	1
5 + 6	20	100	1.84	99	25.53	23.64	1291.4	2.72	5
7 + 8	30	100	1.77	98	28.27	26.50	1493.0	2.47	2
9 + 10	40	100	1.80	97	34.04	32.24	1797.6	2.28	3
11 + 12	50	100	1.80	99	34.56	32.76	1815.8	2.23	1
13 + 14	60	100	1.81	91	32.48	30.67	1693.4	2.32	9
15 + 16	70	100	1.83	32	12.18	10.35	570.7	5.43	68
17 + 18	80	100	1.79	20	9.20	7.47	414.0	6.28	80
19 + 20	90	100	1.82	13	10.34	8.52	474.4	5.63	87

EXPERIMENT TO ESTABLISH THE NUTRITIONAL VALUE OF A FEED CONTAINING ALGAE BUT NO FISH MEAL TO CARP
EXPERIMENT 74/10

The data (Fig. 51) show that the highest average weight was achieved by Feed D, which contained algae but no fish meal. The experiment ran for nearly half a year and there were hardly any losses. It illustrates clearly that carp can be reared without being fed any fish meal. It also shows that a feed without fish meal can equal the performance of a fish meal containing formulation or even — as in this case — better it.

Once the fish meal-less feed was available for carp the influence of this mix on the composition of the carp meat was investigated (Reimers and Meske, 1977). The analysis of whole fish in the four feed groups showed that the carp on the diet containing algae had the lowest dry matter and an extremely low oil content. The analyses are given in Table 7 and illustrated in Fig. 52. The analysis of muscle tissue from the meat showed it to contain only 1.7% oil while the controls on fish meal-containing trout feed gave readings of 5.6%, 6.9% and 6.4% oil respectively (Fig. 53). Oil content is considered in more detail on pp. 120 ff.

The data from this and subsequent experiments confirmed that the feeding of green algae resulted in the by far lowest oil content in fish meat (Meske and Pfeffer, 1978). The same trend was found on analysis of liver and kidneys. It should be noted that the carp on the algae feed mix had a larger gut than the fish not receiving algae (Reimers and Meske, 1977).

TABLE 7

Analysis of whole carp on four feed mixes (percentages)

	Feed A	Feed B	Feed C	Feed D
Dry matter	24.7	30.6	29.0	21.4
Protein	14.7	13.8	15.1	15.0
Oil	12.0	15.0	12.0	4.4
Analysis of meat (muscle fibres) only				
Oil	5.6	6.9	6.4	1.7

When the protein efficiency ratio was calculated, the fish fed algae gave the best conversion values. In the experiments described here a PER of 1.30 was recorded for algae diets (D) and 0.93, 0.94 and 1.28 respectively for the controls A, B and C.

The experiments with grass carp (see below) and carp clearly indicate that feeds containing up to 50% algae produce a significant increase in growth compared to feeds containing fish meal or even without fish meal, as in the SM basic feed mixture series. The boost in growth derived from algae feeding has been noted frequently as an unidentified growth factor for which there is as yet no explanation.

Hepher *et al.* (1978) reported similar results achieved in experiments with warm water table fish fed on diets containing unicellular green algae. They came to the conclusion that algae meal is possibly the only protein source which can be used as a fish meal substitute.

MATERIAL AND METHOD

Duration	180 days
Fish	carp from mating 4 002

Programme	74/10— 1 to 4 50 carp
	74/10— 5 to 8 50 carp
	74/10— 9 to 12 50 carp
	74/10—13 to 16 50 carp

Feed	D *	68% SM3
		32% algae
	A †	trout concentrate
	B †	trout concentrate
	C †	trout concentrate

Feeding regime	*ad lib* for the first 15 days followed by hourly feeding (10 times a day) according to rationing table, p. 68
Tanks	40 l, glass aquaria
Water flow rate	3 l/min
Water temperature	approximately 24°C

* Feed D contained no fish meal. Its formulation was:

Algae (*Scenedesmus*)	32%
Dried whey	34.2%
Soya	26.3%
Fat (melting point 43°C)	6.7%
Vitamin + trace element supplement (0.3% lysine, 0.1% DL-methionine)	

† Feeds A, B and C were conventional trout concentrates containing fish meal

For their composition see Reimers and Meske, 1977.

The analysis of the four feeds used in this experiment is as shown in Table 8 (as % dry matter). The experiment commenced with 200 carp in each feed group (four replications). It

TABLE 8

Feed analyses (percentages)

(Method: Weender)

	Feed A	Feed B	Feed C	Feed D
Protein	40.0	47.9	51.9	38.4
Oil	6.3	6.1	8.6	9.1
Fibre	5.3	4.8	6.3	5.8
Ash	12.6	11.1	9.5	11.4
N-free extracts	35.8	30.1	24.0	35.3

was divided into four time-stages, the last stage being carried out with only 40 fish in each group for reasons of a shortage of facilities (for details see Meske and Pruss, 1977). The summary below gives the results obtained at the end of the first and last stage.

SUMMARY OF DATA AND RESULTS

Stage 1 (days 1 to 83)

Feed		At start		Day 83		Weight increase		FCR
		Number of fish	x Weight (g)	Number of fish	x Weight (g)	(x g)	(%)	
D	68% SM3, 32% algae	200	9.58	198	92.05	82.47	861	1.82
A		200	9.58	196	83.05	73.47	767	1.95
B		200	9.58	197	67.71	58.13	607	2.22
C		200	9.58	197	104.08	94.50	986	1.71

Stage 4 (day 125 to 180 — end of experiment)

Feed	Day 125		Day 180 (end)		Weight increase		FCR
	Number of fish	x Weight (g)	Number of fish	x Weight (g)	(x g)	(%)	
D	40	262.15	40	537.60	275.45	105.1	2.51
A	40	216.53	40	465.05	248.52	114.8	2.38
B	40	178.73	40	347.41	168.68	94.4	2.76
C	40	269.37	40	522.65	253.28	94.0	2.80

THE CONVERSION EFFICIENCY OF CASEIN AND KRILL IN CARP
EXPERIMENTS 76/12 and 77/1

The experiment was carried out to establish whether a feed without fish meal, but using casein and krill as the protein source, can generate satisfactory growth rates in carp. The optimal levels of protein and the component materials supplying the protein were investigated, as was the effect of these various feed formulations on the carp meat.

The krill (*Euphausia superba*) is of particular interest as a feedstuff. It occurs in massive amounts in the Antarctic and is expected to make a considerable contribution to livestock feedstuffs. The krill has a protein content of about 55% (Loerz *et al.*, 1977). Krill contain high levels of fluorine. In a feeding experiment krill containing 2500 mg of fluoride per kg were fed to trout fingerlings. The fluoride content in the muscle tissue of the trout did not increase but concentrations of up to 3100 mg/kg wet weight were found in the bones (Tiews *et al.*, 1981).

The protein content of the feed mixes in this experiment was set at 20, 30 and 40% respectively. The casein inclusion rate was set at 100, 50 and 0% of the total protein. The inverse of these percentages represents the krill inclusion. For example mix 30/50 had a total protein content of 30% of which 50% was casein and 50% krill (Pfeffer and Meske, 1978).

The experiment showed that raising the protein levels does not necessarily achieve better growth rates. As can be seen from the summary of the data, growth and conversion ratio are linked to the proportion of casein versus krill in the mixture. The casein only mixes produced the worst results (FCRs of 3.29, 4.32 and 4.46 respectively). The average weights achieved at the end of the experiment are illustrated in Fig. 54. Mixtures 20/100, 30/100 and 40/100, in which casein was the only source of protein, show the least weight increases. The best weight gains were achieved by mixes in which 50% of the protein was supplied by casein and 50% by krill — i.e. 20/50, 30/50 and 40/50. Of these 50/50 mixes, the 40% protein feed performed best and produced better growth than the fish meal containing control feed A with 54% protein.

In a subsequent experiment 77/1 (detailed in Pfeffer and Meske, 1978) six feed mixtures were tested, all with 40% total protein analysis, made up of varying proportions of casein and krill in 10% increment levels. The mixtures were formulated in the same way as those shown in the experiment 76/12 data table.

Figure 55 shows the average weight of the carp at the end of the experiment after 77 days (weight at start: 35 g). The histogram illustrates clearly the relation between growth and the inclusion rates of casein and krill respectively. Fish exclusively on casein (mix 40/100) performed badly while showing near-continuous improvement in the growth rate as the percentage of the krill component was increased.

Analysis of entire carp bodies at the end of both experiments — 72/12 and 77/1 — demonstrated an increase in their protein content in the dry matter, with (1) rising protein content in the feed; (2) increased krill inclusion. At the same time the oil content diminished with the rise of the percentage of krill in the feed (Pfeffer and Meske, 1978).

MATERIAL AND METHOD

Duration   57 days

Fish   carp from mating 6 026/27

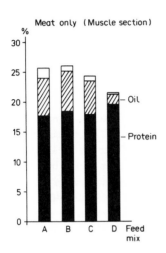

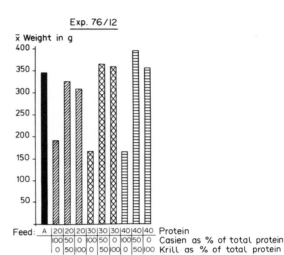

Fig. 53: Analysis of carp meat (section through muscle) for protein and oil as percentage dry matter following the feeding of four different mixtures (Experiment 74/10). D contains no fish meal; A, B and C contain fish meal; hatched columns: oil; solid columns: protein (Reimers and Meske, 1977).

Fig. 54: Fish meal-free feeding of carp with casein and/or krill as the source of protein (Experiment 76/12). Solid columns: control Feed A containing fish meal; hatched columns: feed containing 20% protein; crosshatched columns: feed containing 30% protein; horizontal hatch: feed containing 40% protein.

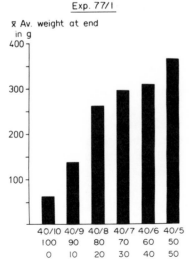

Fig. 55: Feeding of carp with casein and krill as sources of protein (fish meal-free) at various ratios (Experiment 77/1). Average weights at end of experiment. All mixtures contained 40% protein.

Programme   76/12—1, 11, 21   10 carp each (total 30) on feed A
                             76/12—2, 12, 22   10 carp each (total 30) on mixture 20/100

and then continued according to groups and feed mixtures as given in data and results up to

76/12—10, 20, 30   10 carp each (total 30) on mixture 40/0

Feed   feed A: commercial trout concentrate: 54% protein,
                                                  8.3% oil,
                                          11.9% ash

mixtures as below:

Feed code No.*	Casein	Krill	Wheat	Vegetable oil	Vitamin/mineral premix
20/100	20	—	63	10	7
20/50	10	18	55	10	7
20/0	—	36	47	10	7
30/100	30	—	53	10	7
30/50	15	27	41	10	7
30/0	—	54	29	10	7
40/100	40	—	43	10	7
40/50	20	36	27	10	7
40/0	—	72	11	10	7

The heading "Feed components (%)" spans the columns Casein, Krill, Wheat, Vegetable oil, Vitamin/mineral premix.

* The first part of the feed code number refers to the nominal at total crude protein content of a mix. The second part refers to the casein content expressed as a percentage of total protein. The inverse of the casein percentage is then the krill component percentage of total protein.

SUMMARY OF DATA AND RESULTS

Programme code	Feed*	As start ΣWeight (g)	n	x̄ Weight (g)	At end ΣWeight (g)	n	x̄ Weight (g)	Weight gains Σ (g)	%	x̄ (g)	Feed consumed (g)	FCR
76/12–1, 11 + 21	A	2953.7	30	98.46	10 369.7	30	345.66	7416.0	251.07	247.20	10 768	1.45
76/12–2, 12 + 22	20/100	2934.3	30	97.81	4769.3	25	190.77	2324.1	95.05	92.96	7693	3.29
76/12–3, 13 + 23	20/50	2973.0	30	99.10	9241.5	29	324.88	6547.5	227.82	225.78	10 425	1.59
76/12–4. 14 + 24	20/0	2938.0	30	97.60	8645.5	28	308.77	5912.8	216.37	211.17	9620	1.63
76/12–5, 14 + 25	30/100	2936.4	30	97.88	3839.6	23	166.94	1588.4	70.56	69.08	6858	4.32
76/12–6, 16 + 26	30/50	2921.2	30	97.37	10 967.7	30	365.59	8046.5	275.45	268.22	11 617	1.44
76/12–7, 17 + 27	30/0	2972.4	30	99.08	10 801.7	30	366.06	7829.3	263.40	260.95	11 066	1.41
76/12–8, 18 + 28	40/100	2940.0	30	98.00	3817.9	23	166.00	1563.8	69.38	68.00	6973	4.46
76/12–9, 19 + 29	40/50	2911.5	30	97.05	11 545.5	29	398.12	8731.1	310.23	301.07	12 056	1.38
76/12–10, 20 + 30	40/0	2907.0	30	96.90	10 007.9	28	357.43	7294.6	268.85	260.53	10 545	1.45

* First figure denotes total protein in mix. Second figure denotes percentage of casein in protein mix. Its inverse is the krill content of the mix, (i.e. 20/0 contains no casein but 100% of the protein is derived from krill).

CARP (Cyprinus carpio) FEEDING TRIALS WITH FEED FORMULATIONS CONTAINING VARIOUS LEVELS OF PROTEIN AND OIL
EXPERIMENT 80/1

This experiment was carried out in connection with a thesis by Eckhardt (1981) to obtain information on the effect of various protein levels on the growth and body composition of carp. Sixteen feed formulations were used, which combined four levels of oil at 0, 6, 12 and 18% of dry matter (DM) and four levels of protein at 26, 34, 42 and 50% of dry matter.

As Tables 9 and 10 show, the protein (fish meal) was progressively replaced by wheat and the oil (sunflower) by cellulose. Energy was supplied by oil at any given equal protein level in each group. The average feeding rate throughout was 2.8 of liveweight. In addition mixes contained 2.5% gelatine as a pelleting medium.

TABLE 9

Feed formulations (g/kg)

Mix code	Fish meal	Wheat	Sunflower oil	Cellulose	Mineral premix	Vitamins and trace elements
A1	320	454	0	180	36	10
B1	320	454	60	120	36	10
C1	320	454	120	60	36	10
D1	320	454	180	0	36	10
A2	430	356	0	180	24	10
B2	430	356	60	120	24	10
C2	430	356	120	60	24	10
D2	430	356	180	0	24	10
A3	540	258	0	180	12	10
B3	540	258	60	120	12	10
C3	540	258	120	60	12	10
D3	540	258	180	0	12	10
A4	650	160	0	180	0	10
B4	650	160	60	120	0	10
C4	650	160	120	60	0	10
D4	650	160	180	0	0	10

TABLE 10

Analysis of feed mixes* (as percentage of DM)

Mix code	Dry matter (%)	Protein (% DM)	Oil (% DM)	Ash (% DM)	Fibre (% DM)	Nfe** (%%)	Energy [†] (KJ/g DM)
A1	96.95	29.1	0.3	7.0	15.8	45.1	14.26
B1	96.44	28.5	6.6	6.7	12.6	45.6	17.33
C1	97.08	28.5	12.5	6.7	6.1	46.2	19.46
D1	96.51	27.3	19.1	6.8	1.0	45.8	21.84
A2	96.85	37.0	0.5	7.3	16.8	38.4	15.66
B2	96.80	35.1	6.6	7.1	11.5	39.7	17.75
C2	97.29	35.5	13.1	7.2	6.2	38.1	20.09
D2	96.67	36.3	19.5	7.4	1.1	35.7	22.60
A3	96.44	45.5	0.6	7.8	16.5	29.6	15.96
B3	95.54	46.2	6.8	7.8	11.5	27.7	17.96
C3	97.29	42.9	13.5	7.8	7.6	28.2	20.43
D3	96.83	44.8	19.6	7.9	1.2	26.5	22.97
A4	95.72	54.8	0.7	8.5	16.8	19.2	16.84
B4	97.14	52.2	7.0	8.5	11.4	20.9	18.96
C4	97.59	49.9	13.6	8.4	6.5	21.6	21.35
D4	97.43	54.4	19.5	8.4	1.1	18.6	23.68

*   Method of analysis: Weender
†   Cellulose
** Nitrogen-free extracts
DM = dry matter

Figure 56 and the data and results show that the best growth rate was achieved by the 42% protein group of mixes (A3, B3, C3 and D3). Mixes A1 to A4 contained no oil, B1 to B4 contained 6%, C1 to C4 12% and D1 to D4 18% oil. Overall the growth of the carp improved with increasing oil content, and this is more pronounced at low protein inclusions.

When the tissues of fish on the high-oil and low-protein mixes were analysed, a notable increase in their oil content was found. The mix with the 18% oil and 26% protein resulted in a body oil content of nearly 16%. In contrast the mix with the same oil of 18% but containing

50% protein gave a body oil analysis of 9.27%. The best results in terms of feed conversion, body protein and oil content and productive protein value (PPV) were achieved by mix C3 which had a content (as percentage of dry matter) of 42% protein and 12% oil.

MATERIAL AND METHOD

Duration	70 days			
Fish	carp X 9010			
Programme	80/1—	1 + 17	fed mix	A1
		2 + 18		A2
		3 + 19		A3
		4 + 20		A4
		5 + 21		B1
		6 + 22		B2
		7 + 23		B3
		8 + 24		B4
		9 + 25		C1
		10 + 26		C2
		11 + 27		C3
		12 + 28		C4
		13 + 29		D1
		14 + 30		D2
		15 + 31		D3
		16 + 32		D4

No. of fish per tank	10	
No. of fish per test	20	
Feeding frequency	9 times daily	
Feeding rate	Mean liveweight (g)	Feed fed (percentage liveweight per day)
	30— 50	5
	50—100	4
	100—400	3.5
Tanks	40 l, glass aquaria	
Water temperature	mean 23.8°C	
Water flow rate	mean 2.8 l/min	
Weighing	weekly	

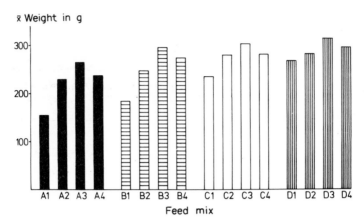

Fig. 56: Average weight of the carp at the end of test of 16 feed mixes, A1 to D4 (Experiment 80/1).

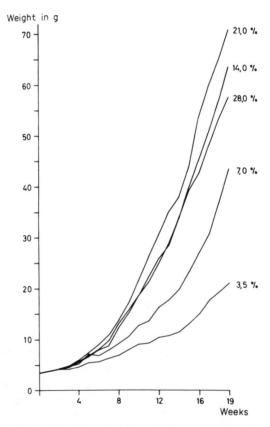

Fig. 57: The effect of daily rationing levels (Experiment 67/71). Average growth rate of carp on daily rations of 3.5%, 7%, 14%, 21% and 28% respectively of liveweight.

DATA AND RESULTS

| Feed mix code | Feed analysis | | Weight and number of fish | | | | FCR | Growth : analysis of tissue composition as percentage of liveweight | | | PPV* |
| | Protein (%) | Oil (%) | at start | | at end | | | | | | |
			No.	Mean weight (g)	No.	Mean weight (g)		DM	Prot.	Oil	(%)
A1	0	26	20	42.12	20	155.97	1.73	19.62	13.82	1.02	27.71
A2	0	34	20	41.31	20	229.11	1.37	20.19	14.56	1.53	29.59
A3	0	42	20	41.32	20	264.98	1.29	19.60	14.59	1.26	25.58
A4	0	50	20	41.00	20	237.95	1.35	18.94	14.12	0.79	20.01
B1	6	26	20	40.94	20	183.77	1.60	22.55	13.59	3.93	30.73
B2	6	34	20	41.32	20	246.45	1.31	21.69	13.66	3.24	30.63
B3	6	42	20	41.31	20	269.99	1.24	23.07	14.42	4.29	26.32
B4	6	50	20	40.96	20	273.62	1.35	22.81	14.76	3.84	21.45
C1	12	26	20	41.17	20	234.07	1.35	29.41	12.96	11.13	34.78
C2	12	34	20	41.39	20	279.58	1.32	27.43	13.29	9.06	29.31
C3	12	42	20	41.11	20	303.33	1.24	26.22	14.00	7.78	27.35
C4	12	50	20	41.24	20	281.82	1.35	25.44	14.40	6.72	21.95
D1	18	26	20	41.26	20	268.40	1.29	32.84	11.81	15.64	34.78
D2	18	34	20	40.89	20	282.03	1.36	30.29	13.26	12.23	27.71
D3	18	42	20	41.21	20	314.42	1.25	28.18	13.91	9.31	25.70
D4	18	50	20	41.20	20	297.22	1.31	27.41	13.79	9.27	20.65

* PPV : Productive protein value
Source : Meske and Becker, 1981

FA-I

THE INFLUENCE OF FEEDSTUFFS ON THE PROTEIN AND OIL CONTENT OF CARP MEAT

The experiments described above dealt specifically with the influence of feed formulations containing the same components in different proportions on carp growth and conversion efficiency.

By conducting a large number of experiments during recent years it has been possible to obtain comprehensive information on the quality of carp meat as a function of the source and amounts of the proteins in the feed (Meske and Pfeffer, 1978). The feeding experiments were carried out at Ahrensburg. The fish meat analyses were undertaken at the Institute of Animal Nutrition at Goettingen University.

It seemed pertinent to establish first what changes in the meat or the body composition followed starvation. For this carp of various weights were kept without feeding for 28 days at 23°C in flow through aquaria fed by spring water.

Some of the carp were killed and their meat analysed before the start of the experiment. During the 28 days when they were not fed the carp lost between 11% and 16% of their weight. The heavier fish in the group — at about 900 g individual weight — lost least in calorific value, comprising about 50 g each of protein and oil. The smaller fish in the group lost about 50 g each, nearly half their oil content, but showed little change in their protein levels compared to the controls (Meske and Pfeffer, 1978; Nijkamp et al., 1974). In experiments with longer starvation periods it was found that the oil content of the meat was reduced from about 54 g/kg to 29 g/kg after 159 days. The protein, on the other hand, only declined by 4 g/kg. Altogether the carp in this experiment lost 40% of their weight, the average weight being reduced from 403 to 243 g (Pfeffer et al., 1977).

Experiments by Huisman (1976) and also by Kausch and Ballion-Cusmano (1976) show that the rate of feeding intensity — defined as daily amounts fed as percentage of weight — has no bearing on the protein content of carp meat. There is practically no difference in the body protein content as the result of feeding 1—9% of bodyweight daily. The same applies to ash. The oil content, however, increased considerably. At the same time the feed conversion ratio deteriorated as the rate of feeding increased.

The composition of the carp meat was not only influenced by the amounts fed. The decisive factor was the feedstuffs used. The protein and oil contents of carp are of some importance. It was found that the protein could be raised by as much as 15% by varying the protein source in the feed and not by raising its protein content. The experiments which showed this most clearly employed feed formulations containing between 35% and 53% protein, the origin of the protein ranging over fish meal, algae, whey and casein. The carp was found to contain extremely low levels of oil when on feeds which did not include fish meal. On a diet without fish meal the oil content sunk to between 1.3% and 4.8% compared with 8.5% oil in fish fed on trout concentrate which included fish meal (Meske and Pfeffer, 1977). (See p. 108 ff.)

The protein level in carp can be varied considerably and was influenced by feeding mixtures with the same protein analysis but differing in the source of protein. This was shown in the results of a series of experiments in which the protein sources were casein and krill, fish meal being excluded (Pfeffer and Meske, 1978). (See p. 111 ff.)

The experiments showed that increasing the protein in the feed from 20% to 30% and then to 40% gave no discernible increase in the carp's body protein when krill was the feed protein source. With casein as the protein source in the feed, the carp's body protein decreased significantly with rising inclusion rates. The oil content in the fish always decreased as protein content increased.

Experiment 77/1 — see above — confirms that the difference in the body composition of the carp is determined by the sources of the protein in the feed. In this experiment the feeds had a 40% protein analysis, the protein being supplied exclusively by casein and krill. As the proportion of krill increased from 0% to 50% — as g of nitrogen — so the food conversion ratio improved from 4.65 to 1.15 (see Fig. 53). The corresponding rise in the protein content of carp meat — from 10.6% to 14.2% — and the decline of the oil content is also remarkable. This issue is somewhat complex and further experiments are being conducted to clarify it and allow a proper appraisal of the results.

It is equally important to establish the requirement and use of the mineral trace elements. Comprehensive data on minerals and their influence on the growth of warm-blooded farm animals are available (*viz.* Guenther, 1972). In the case of fish the knowledge of this aspect of nutrition is comparatively modest but clearly indicates that the mineral requirements of warm-blooded animals bear little or no relation to the needs of table fish. This has become noticeable from experiments conducted in aquaria during recent years when fish were fed analytically determined feed mixtures (Pfeffer *et al.*, 1977; Pfeffer and Potthast, 1977; Meske and Pfeffer, 1977).

Pfeffer's work (1978) has established exact data on the ash content of whole carp, as well as that of the different parts of the carp. They include values for calcium, phosphorus, magnesium, potassium and sodium. The results show, for instance, that the fish in the experiments had a higher phosphorus content than is found in pigs or in cattle. The fish bones, in contrast, contained less calcium than the bones of mammals. Feeding experiments and analyses were undertaken to establish the requirement for mineral trace elements in the carp (Pfeffer and Meske, 1979). Feeds were fortified with various levels of minerals in the form of fish meal ash. However, no clear conclusions could be reached in respect of the effect of mineral levels in feeds. Frenzel and Pfeffer demonstrated in 1982 that fish can absorb some trace elements from the water.

THE EFFECT OF DAILY RATIONING LEVELS
EXPERIMENT 67/71

This experiment involved 600 young carp with an average weight of 3.4 g at the start. It demonstrated the dependence of growth on the amount of feed given daily, and led to an interesting result: within the parameters of this experiment a daily ration of 28% of liveweight proved excessive and resulted in less growth than daily rations of 21% and 14% of liveweight (Fig. 57). Evidently not all the feed offered was consumed, causing deterioration of the water quality and reduction in the oxygen level — although water was circulating through the tanks at a rate of 4 l/min.

The 28% of liveweight daily ration gave a statistically confirmed poorer rate of growth in carp than a 21% ration. At the 28% feeding rate the fatalities occurring during the experiment were also the heaviest.

In practice, attention must be paid to the relationship between the amount of pellets fed daily, the tank size, the water flow rate and the oxygen content of the water.

MATERIAL AND METHOD

Duration	131 days
Fish	carp from mating 7025
Programme	67/71—1 and 6: 60 carp each daily ration: 3.5% of liveweight
	67/71—2 and 7: 60 carp each daily ration: 7% of liveweight
	67/71—3 and 8: 60 carp each daily ration: 14% of liveweight
	67/71—4 and 9: 60 carp each daily ration: 21% of liveweight
	67/71—5 and 10: 60 carp each daily ration: 28% of liveweight
Feed	trout diet A hourly between 7.30 and 17.00 hours
Tank	40 l, glass
Water flow rate	4 l/min

DATA AND RESULTS

	67/71—1 & 6 Ration: 3.5% of live weight daily	67/71—2 & 7 Ration: 7.0% of live weight daily	67/71—3 & 8 Ration: 14.0% of live weight daily	67/71—4 & 9 Ration: 21% of live weight daily	67/71—5 & 10 Ration: 28% of live weight daily
At start					
Number	120	120	120	120	120
Weight (g)	410	410	410	410	410
$\bar{x}$ Weigh (g)	3.4	3.4	3.4	3.4	3.4
At end					
Number	103	104	95	98	87
Weight (g)	2165	4510	6030	6945	5030
$\bar{x}$ Weight (g)	21.0	43.4	63.5	70.9	57.8
Weight increase (g)	1813	4155	5705	6610	4733
Weight increase (%)	515.1	1170.4	1755.4	1973.1	1593.6
$\bar{x}$ Weight increase (g)	17.6	40.0	60.1	67.4	54.4
Food fed (g)	4105	9955	24 844	37 930	39 320
CR	2.26	2.40	4.35	5.74	8.31

THE EFFECT OF DAILY RATIONING LEVELS
EXPERIMENT 67/7

Experiment 67/7 confirms the results of other experiments. Although there were only three fish in each group, it did run over 11 months. Statistically significant results agree almost entirely with those of experiment 66/2. Figure 58 shows clearly the divergencies in the growth curves over 48 weeks. In contrast to experiment 66/2 a daily ration of 5% of liveweight differed statistically in this experiment from a 4% daily ration. This experiment was carried out in co-operation with B. Luehr.

MATERIAL AND METHOD

Duration	336 days
Fish	carp from batch 12 66
Programme	67/7—1: 3 carp Ration: 2% of liveweight daily
	67/7—2: 3 carp Ration: 3% of liveweight daily
	67/7—3: 3 carp Ration: 4% of liveweight daily
	67/7—4: 3 carp Ration: 5% of liveweight daily
Feed	trout diet A
Tank	40 l, plastic
Water flow rate	2.5 l/min

DATA AND RESULTS

	67/7—1 Ration: 2% of liveweight daily	67/7—2 Ration: 3% of liveweight daily	67/7—3 Ration: 4% of liveweight daily	67/7—4 Ration: 5% of liveweight daily
At start				
Number	3	3	3	3
Weight (g)	130	130	130	130
$\bar{x}$ Weight (g)	43.3	43.3	43.3	43.3
At end				
Number	3	3	3	3
Weight (g)	1610	1855	3125	4500
$\bar{x}$ Weight (g)	536.7	618.3	1041.7	1500
Weight increase (g)	1480	1725	2995	4370
Weight increase (%)	1138.5	1326.9	2303.9	3361.5
$\bar{x}$ Weight increase (g)	493.3	575	998.3	1456.7
Food fed (g)	3985	5080	10 660	20 005
CR	2.69	2.94	3.56	4.58

THE EFFECT OF DAILY RATIONING LEVELS (FEEDING INTENSITY)
EXPERIMENT 81/11

The ration scale for carp given on p. 41 and reproduced below has been successfully used at Ahrensburg for many years. However, as feeding is normally manual, dispensing is restricted to the working day of 9 hours, from 07.30 hours to 16.30 hours.

TABLE 11

The Ahrensburg scale of rationing (= 100%)

Mean weight of fish (g)	Amount of as percentage of liveweight per day (%)	Feeding frequency
10	10	hourly
10 – 20	8	hourly
20 – 30	6	hourly
30 – 50	5	hourly
50 – 100	4	hourly
100 – 400	3.5	hourly
400 – 600	3	hourly
600 – 1000	2.5	2-hourly
over 1000	2	2-hourly

Using electrically controlled feeders in artificially lit darkrooms (see p. 84 for description) it became possible to extend the daily feeding period to 17 hours. This experiment was conducted to test the possibility of a bigger daily ration — or higher feeding intensity — resulting in correspondingly greater weight gains. Taking Table 11 as 100% feed intensity, the feeding scale was raised to 125%, 150% and 175% respectively.

From the results, which are illustrated in Fig. 59, it can be seen that the higher feeding scales only achieved modest success. During the 5 months of the experiment the carp in the 125% feeding intensity group clearly had a better growth rate than the 100% group, but their feed conversion efficiency deteriorated from 2.16 to 2.41. The higher feeding levels of 150% and 175% showed no advantage, and the feed conversions of 3.05 and 3.32 indicate that the amounts of feed dispensed were too large even though the feeding time was extended to 17 hours each day.

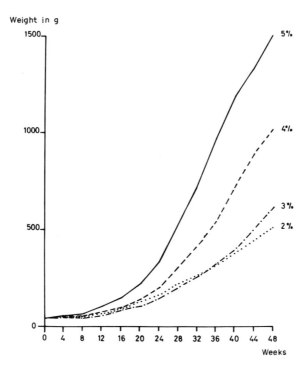

Fig. 58: Effect of daily ration levels (Experiment 67/7). Growth of carp on rationing levels of 2%, 3%, 4% and 5% on liveweight respectively.

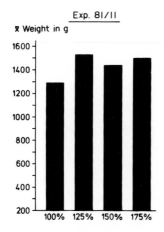

Fig. 59: The effect of daily rationing levels (feeding intensity) (Experiment 81/11). 100% = scale of rationing as above.

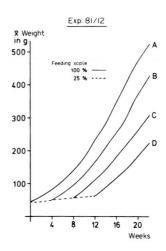

Fig. 60: Growth retardation (Experiment 81/12). Growth of four groups of carp by change of feeding intensity (Meske, 1983).

MATERIAL AND METHOD

Duration	153 days
Fish	carp (X 0009)
Programme	81/11—1 + 2; rationed at 100% of Ahrensburg scale
	81/11—3 + 4; rationed at 125% of Ahrensburg scale
	81/11—5 + 6; rationed at 150% of Ahrensburg scale
	81/11—7 + 8; rationed at 175% of Ahrensburg scale
	15 fish per tank
	30 fish per treatment
Lighting	two 40 W strips for 17 hours each day commencing at 07.30 hours
Tanks	400 l (1 × 1 m base) GRF
Water temperature	$24.3 \pm 0.92°C$
Water flow rate	17 l/min approximately
Feed	trout concentrate. Analysis: protein: 47%, oil: 8%, fibre: 3.5%, ash: 10%
Feeding frequency	9 times daily at 2-hourly intervals by automatic feeder
Weighing	weekly

DATA AND RESULTS

Experiment code No.	Feeding scale* (%)	No.	Fish at start Total weight (g)	Mean weight (g)	No.	Fish at end Total weight (g)	Mean weight (g)	Weight increase Total (g)	(%)	Feed dispensed (g)	FCR
1 + 2	100	30	6021	200.7	29	37 419.3	1290.3	31 398.3	521.48	67 971	2.16
3 + 4	125	30	6023	200.8	29	44 382.4	1530.4	38 359.4	636.88	92 352	2.41
5 + 6	150	30	6024	200.8	30	43 253.6	1441.8	37 229.6	618.02	113 671	3.05
7 + 8	175	30	6022	200.7	30	44 925.2	1497.5	38 903.2	646.02	129.286	3.32

* Ahrensburg feeding scale = 100%

GROWTH RETARDATION
EXPERIMENT 81/12

Research has demonstrated that inadequate environmental conditions retard growth in fish and that on qualitative or quantitative improvement in the environment, fish display compensatory growth (e.g. Langhans and Schreiter, 1928; Walter, 1931; Mann, 1960; Miaczynski and Rudzinski, 1961).

This experiment was carried out to observe the effect of low—high changes in the feed availability, the environment remaining constant. All fish in the experiment were fed on the standard Ahrensburg feed scale (AFS) (100%) as given on page 124 and then divided into four groups:

Group A: fed on the Ahrensburg feeding scale throughout (100%);
Group B: was fed 25% of the AFS for the first 4 weeks of the experiment and then at the full AFS (100%) rate;
Group C: was fed 25% of the AFS for the first 8 weeks of the experiment followed by the full AFS rate;
Group D: was fed 25% of the AFS for the first 12 weeks of the experiment and then put on the full AFS regime.

Figure 60 illustrates the growth of the fish in the four groups. Group A on Ahrensburg feeding scale throughout the 152 days of the experiment achieved the best and fastest growth. Groups B, C and D grew slower to less weight than the control Group A fish fed continuously on the Ahrensburg scale during the time of the experiment.

The results of this experiment indicate that there is no compensatory growth during an adequate feeding period if it follows initial growth retardation due to underfeeding. It also demonstrates that deliberate growth retardation is not necessarily economically advantageous in warm water installations and that continuous and adequate feeding levels achieves the best results.

MATERIAL AND METHOD

Duration	152 days	
Fish	carp X 1001	
Programme	81/12—1 to 5:	60 carp on full (100%) feed scale throughout
	81/12—6 to 8:	36 carp on 25% of feed scale for first 4 weeks of experiment. After that on full scale
	81/12—9 to 11:	36 carp on 25% of feed scale for first 8 weeks of experiment. After that on full scale
	81/12—12 to 14:	36 carp on 25% of feed scale for first 12 weeks of experiment. After that on full scale
Feed		trout concentrate. Analysis: protein: 47%, oil: 8%, fibre: 3.5%, ash: 10%

Feed regime	feed weighed out daily and divided into 5 portions per day for fish on 100% scale and 3 portions per day for fish on 25% scale
Feed scale	on 100% scale
	fish mean weight: 30— 50 g: 5% of liveweight/day
	50—100 g: 4% of liveweight/day
	100—400 g: 3.5% of liveweight/day
Weighing	weekly
Tank	40 l, glass aquaria
	at start each aquarium contained 12 fish which were transferred into 2 aquaria containing 6 fish each when the average weight of the fish reached 280 g
Water temperature	23°C (mean)
Water flow rate	3.5 l/min (mean)

DATA AND RESULTS

Experiment code No.	Feeding scale rate	Fish at start			Fish at end			Weight increase		Feed fed	FCR
		No.	Total Weight (g)	Mean Weight (g)	No.	Total Weight (g)	Mean Weight (g)	Total (g)	(%)	(g)	
1— 5	100% through-out	60	2730.1	45.50	58	30 337.0	523.00	28 044.9	1027.25	59 070.6	2.11
6— 8	25% for first 4 weeks; there-after 100%	36	1639.2	45.53	36	15 285.0	424.58	13 645.8	832.47	24 442.6	2.01
9—11	25% for first 8 weeks; there-after 100%	36	1637.9	45.50	34	10 448.0	307.29	9167.1	559.69	17 882.4	1.95
12—14	25% for first 12 weeks; there-after 100%	36	1639.6	45.54	36	8015.0	222.62	6375.4	388.84	11 401.9	1.79

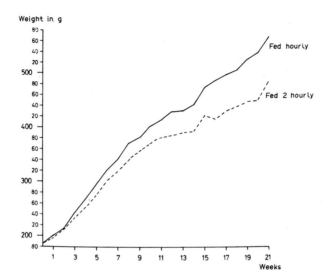

Fig. 61: The effect of daily feeding frequency (Experiment 66/100). Growth rate with hourly feeding shown by solid line and 2-hourly by broken line. Both on same amount of feed; 20 carp in each of 4 tanks.

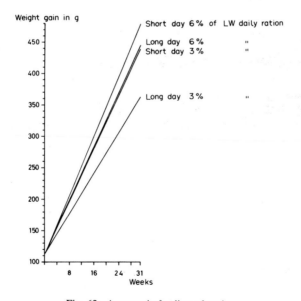

Fig. 62: Automatic feeding of variable duration and ration levels (Experiment 66/1). Growth of four groups of 40 carp each on two levels of feed rationing at hourly and ½-hourly intervals. The 6% short-day group was fed hourly. All other groups ½-hourly.

THE INFLUENCE OF DAILY FEEDING FREQUENCY
EXPERIMENT 66/100

The frequency of daily feeding is of major importance in intensive warm water culture because the fish are, in contrast to pond culture, entirely dependent on the nutrition provided for them.

Experiment 66/100, which involved 40 fish for an experimental period of over 5 months, examined growth rate with hourly and 2-hourly feeding intervals. The daily ration for all fish is shown in the tables on pages 68 and 124. The higher feeding frequency was beneficial to growth rate, as can be seen from Fig. 61. On average fish fed hourly increased their weight by 430.2 g, while those fed 2-hourly only put on 337.3 g over the 5 month period.

In warm water cultures the more frequent feeding with correspondingly smaller amounts must therefore be given preference over other methods. This should be considered, for instance, when designing automatic feeders.

MATERIAL AND METHOD

Duration	155 days
Fish	carp from batch 6 65
Programme	66/100—1 and 2: 10 carp each fed hourly
	66/100—3 and 4: 10 carp each fed 2-hourly
Feed	trout diet A
Tank	40 l, plastic
Water flow rate	2.5 l/min

DATA AND RESULTS

	66/100—1 and 2 fed hourly	66/100—3 and 4 fed 2-hourly
At start		
Number	20	20
Weight (g)	3745	3715
$\bar{x}$ Weight (g)	187.3	185.8
At end		
Number	18	20
Weight (g)	11 115	10 460
$\bar{x}$ Weight (g)	617.5	523.0
Weight increase (g)	7744	6745
Weight increase (%)	229.7	181.6
$\bar{x}$ Weight increase (g)	430.2	337.3
Food fed	34 300	33 522
CR	4.43	4.97

AUTOMATIC FEEDING OF VARIABLE DURATION AND RATION LEVELS
EXPERIMENT 66/1

Experiment 66/1 combines an investigation into the effect of daily feeding (long day, short day), the frequency of feeding (hourly and ½-hourly respectively) and rationing (3% and 6% of liveweight, respectively, daily).

160 carp of equal weight at the start of the experiment were fed from an electrically powered automatic feeder controlled by a time switch. There were four groups of carp on different feeding regimes which are set out in the data below.

Figure 62 illustrates the results of the experiment. The short-day group was fed at ½-hourly intervals between 07.30 and 17.00 hours at a total daily rate of 6% of their weight. This group achieved the best growth of 365 g per fish (tank 1—7 and 1—8). The worst performance came from fish fed 3% of their weight daily during 15 day-time hours at hourly intervals. Their average individual weight gain was 248 g (tanks 1—1 and 1—2).

MATERIAL AND METHOD

Duration	220 days
Fish	carp from batch 665
Method	automatic feeders were suspended above the aquaria; a measured quantity of feed was dispensed by timeswitch controlled electric motor.
	66/1—1 and 2: 20 carp each. Feeding: hourly from 5.00 to 20.00 hours (15 h) Ration: 3% of liveweight daily
	66/1—3: 20 carp each. Feeding: hourly from 7.30 to 17.00 hours (9½ h) Ration: 3% of liveweight daily
	66/1—5 and 6: 20 carp each. Feeding: hourly from 5.00 to 20.00 hours (15 h) Ration: 6% of liveweight daily
	66/1—7 and 8: 20 carp each. Feeding: ½-hourly from 7.30 to 17.00 hours (9½ h) Ration: 6% of liveweight daily
Feed	trout diet A
Tank	40 l, plastic
Water flow rate	2.5 l/min

DATA AND RESULTS

	66/1—1 and 2 Feeding: hourly 5.00 to 20.00 hrs Ration: 3% of LW daily	66/1—3 and 4 Feeding: hourly 7.30 to 17.00 hrs Ration: 3% of LW daily	66/1—5 and 6 Feeding: hourly 5.00 to 20.00 hrs Ration: 6% of LW daily	66/1—7 and 8 Feeding: ½ hourly 7.30 to 17.00 hrs Ration: 6% of LW daily
At start				
Number	40	40	40	40
Weight (g)	4570	4575	4570	4575
$\bar{x}$ Weight (g)	114.2	114.4	114.2	114.4
At end				
Number	31	29	24	28
Weight (g)	11 235	12 740	10 690	13 425
$\bar{x}$ Weight (g)	362.4	439.3	445.4	479.3
Weight increase (g)	7689	9422	7949	10 222
Weight increase (%)	216.8	284.0	290	319.1
$\bar{x}$ Weight increase (g)	248.0	324.9	331.2	365.0
Food fed (g)	42 220	55 435	70 065	85 640
FCR	6.01	5.88	8.81	8.34

LW = liveweight

AUTOMATIC ON DEMAND FEEDING
EXPERIMENT 67/29

The *ad lib* – or demand – feeder shown in Fig. 63 was constructed by O. Cellarius at Ahrensburg. Within a few days the fish get used to the rubber bait which operates the hopper flap. In contrast to the commercially available equipment where it is only necessary to touch a pendulum, the Cellarius feeder only releases feed when the bait is actually pulled. No food is dispensed when it is accidentally touched.

Experiment 67/29 was conducted to see whether carp could be fed exclusively over long periods on the demand system and whether they fed more actively during the day or the night. During a period of nearly 9 months 54 carp were fed by this self-feeder only. They gained an average of 256.3% in weight.

The demand feeder was made available to three groups of carp (tanks 1—3) for 24 hours a day, to another three groups (tanks 4—6) during the day only (07.30 to 17.00 hours) and one group (tanks 7—9) during the night only (17.00 to 07.30 hours). Feed consumption of the day-

and night-fed fish was almost the same for all groups. It differed by less than 2 kg over the period of the experiment.

However the fish on the 24 hour feed availability regime had a considerably higher feed consumption, as can be seen from data and results. Their feed conversion ratio was the worst at 7.1 while the daily-fed fish converted at 4.13 and the fish at night at 4.45.

Figure 64 illustrates feed consumption and conversion during the first 18 weeks of the experiment. The groups displayed insignificant differences in their growth during this time span (Fig. 65).

It is of interest to note the gradual decrease of daily feed intake with increasing weight (Fig. 66). This can be taken as an indicator of the actual feed requirement of the fish within the described experimental parameters.

MATERIAL AND METHOD

Duration	265 days
Fish	carp from batch 5 66
Method	67/29—1 to 3: 6 carp each, fed on demand for 24 hours/day
	67/29—4 to 6: 6 carp each, feed offered 7.30—17.00 hours
	67/29—7 to 9: 6 carp each, feed offered 17.00—7.30 hours
	(for description of feeder see Fig. 63)
Feed	trout diet A
Tank	40 l, plastic
Water flow rate	2.5 l/min

DATA AND RESULTS

	67/29—1 to 3 24 hours on demand	67/29—4 to 6 feed offered 7.30—17.00 hours	67/29—7 to 9 feed offered 17.00—7.30 hours
At start			
Number	18	18	18
Weight (g)	5855	5840	5895
$\bar{x}$ Weight (g)	325.3	324.4	327.5
At end			
Number	18	18	17
Weight (g)	20 480	21 440	19 595
$\bar{x}$ Weight (g)	1137.8	1191.1	1152.6
Weight increase (g)	14 625	15 600	14 027
Weight increase (%)	249.8	267.1	251.9
$\bar{x}$ Weight increase (g)	812.5	866.7	825.1
Food eaten (g)	102 465	64 350	62 430
CR	7.01	4.13	4.45

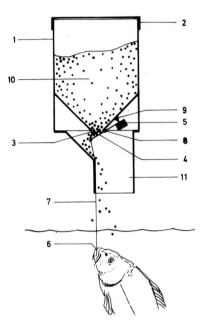

Fig. 63: The Cellarius self-feeder. Hopper (1) with lid (2) has opening at the bottom (3) which is kept closed by flap (4) and counterweight (5). To feed, the fish has to pull the rubber bait (6) which, through a cord (7), opens flap (4) until the hinge (8) and weight (5) hit the stop (9). As long as the fish pulls, feed (10) drops through the opening (3) into the water. Chute (11) protects the mechanism against splashing. Counterweight (5) Shuts opening (3) immediately the fish stops pulling.

Fig. 64: Automatic on-demand feeding in carp (Experiment 67/29). Histogram covers the first 18 weeks of the experiment and shows feed take-up and conversion ratio.

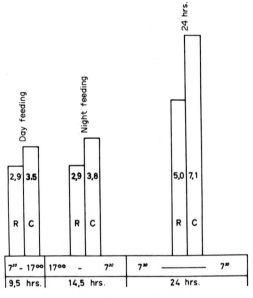

R : Daily feed dispensed as percentage of live weight

C : Feed conversion ratio

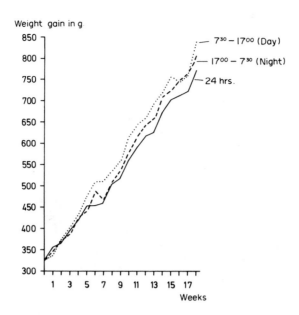

Fig. 65: Automatic on-demand feeding in carp (Experiment 67/29). Growth in three groups of 18 carp each on-demand feeders during the first 18 weeks of the experiment. ———— Feed available for 24 hours a day; . . . . . . feed available between 07.30 and 17.00 hours (day); – – – – – feed available between 17.00 and 07.30 hours (night).

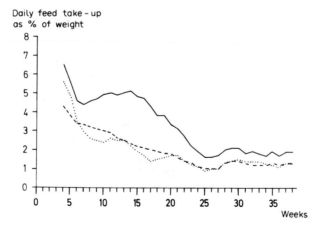

Fig. 66: Demand feeding in carp (Experiment 67/29). Change in daily feed uptake as percentage of liveweight. Calculated on sliding average. ———— Feed available 24 hours a day; . . . . . . feed available from 07.30 to 17.00 hours (9½ hours); – – – – – feed available from 17.00 to 07.30 hours (14½ hours).

SUMMARY OF THE FEEDING EXPERIMENTS

The experiments on the feeding of carp fry show clearly that trout concentrate is not a suitable feed for this purpose. The feeding of live food cannot as yet be avoided for carp fry as compounders have so far not produced a formula which can be fed satisfactorily after the yolk sac is absorbed.

Artemia larvae have the advantage that they can be used universally. Algae are not a complete feed for fry but they can be used at a later stage when they are of practical value because they result in better growth than trout concentrate. It was not possible to ascertain the optimal feed formulation under the available experimental conditions.

Feed quantity, of any specific feed, as well as feeding frequency, is a strong determinant for carp growth. Both these factors must be balanced carefully for optimal economic results in relation to growth rate and feed conversion ratio. Optimal amounts of feed and feeding frequency are dependent on the age of the fish, feed quality, water temperature, water flow rate and other considerations. A demand feeder gives the fish a chance to obtain at all times the food it requires, provided the factors listed above are taken into consideration. The fish is a better judge of its environment than any measuring device.

BREEDING

In this section experiments in breeding are described: the shoot carp or tobi koi phenomenon is examined and techniques for breeding programmes are discussed. The latter includes a method of marking brood fish, progeny testing at Ahrensburg and work on the breeding out of intramuscular bones in carp.

*TOBI-KOI OR SHOOT CARP**
EXPERIMENT 68/33

Shooting is a widespread phenomenon among fish species. It is regularly observed in pond carp populations. Experiment 68/33 was conducted to ascertain whether shooting had a genetic cause or is due to environmental factors.

---

* Editor's note: The tobi-koi or shoot carp phenomenon was first reported by Japanese workers (Nakamura and Kasahara, 1955, 1956, 1957, 1961). It was subsequently investigated by Israeli workers (Wohlfahrt and Moav) — among others — who called these fish koftsim or jumpers. Other workers, some working on other species (i.e. catfish: Konikoff and Lewis, 1974) often gave other descriptive names.

The phenomenon can be observed after 20 days and can be described as an asymmetrical increase in weight variation in a carp population due to interactions between individuals in the population and resulting from competition for food.

A full survey of the subject, which includes the translation of the three original classical Japanese papers and a comprehensive list of references can be found in Wohlfahrt, G.N., 1977: *Bamidgeh*, Vol. 29 (1977), No. 2 (IS OSSN 0005 — 4577), pp. 35—56, published jointly by the Fisheries Department of the Israeli Ministry of Agriculture and the Fish Breeders Association, Nir-David 1950, Israel.

Wohlfahrt concludes that quantity and quality of feed are the primary factors of shooting, and that

One hundred full sibling carps, hatched on 7 January 1968 and having a mean weight of 51 g on 29 April 1968, were reared in an aquarium. On 3 of May the five heaviest and the five lightest individuals were selected from these 100 fish and each of the two groups placed in separate tanks. As can be seen from Fig. 67 and the data, considerable differences in weight developed between the two groups within the 9 months of the experiment's duration. The heaviest fish of the originally lightest carp group caught up with the lightest carp of the original heavy group after about 6 months of the experiment on 1 November 1968. Up to the end of the experiment it grew 105 g more.

The course of the experiment permits the interpretation that the differences in weight are due to behavioural factors. If, in particular, the growth of the biggest fish among the smaller carp group at the start of the experiment is noted, genetic causes for shooting (i.e. early maturity or precocity) must be excluded, as otherwise this would have been observable before the start of the experiment. The environmental conditions were the same for all fish, as was the feed.

It is possible that a dominant hierarchy — a pecking order — prevails within the tank. Fish which may have remained small because of the hierarchy developed unchecked once they were isolated from the higher-ranking bigger fish. Indeed each group of five either retained its ranking or developed a new one. The same phenomenon appears to occur within the original five heaviest carp group. Within this group the lowest-ranking fish lagged behind in growth, so that at the end of the experiment it weighed less than the heaviest fish of the original lightweight group.

MATERIAL AND METHOD

Duration	274 days
Fish	carp from batch 8002
Method	100 carp were taken from an aquarium and used as follows:
	68/33—1: the 5 heaviest carp
	68/33—2: the 5 lightest carp
Feed	trout diet A
Tank	40 l, plastic
Water flow rate	2.5 l/min

---

stocking density has no clear influence unless it leads to reduced feed availability, particularly during the nursing stages.

This footnote is included because there is no carp farming on any significant scale in the British Isles and British fishery biologists and fish farmers consequently may be unfamiliar with the phenomenon or its terminology. Perhaps precosity could be a useful English term for it.

Although the fish appear all to have been F_1 full-sibs there are no indications that they were back-crossed to see whether the shooting is not a possible case of heterozygosity.

See also page 144, paragraph 6.

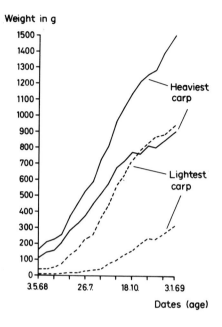

Fig. 67: Shoot carp (Experiment 68/33). Growth rate of the heaviest and lightest carp representing extreme weights within a group of 100 full siblings. Graph stops 5 weeks before end of experiment.

Fig. 68: Investigation of the shoot carp phenomenon (Experiment 69/5). Growth of five carp of the same weight at the start of the experiment (tank 69/5–2) over 18 weeks.

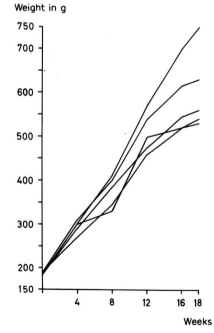

DATA AND RESULTS

		68/33−1 5 heaviest carp	68/33−2 5 lightest carp
At start			
Weight (g)	heaviest carp	265	40
	lightest carp	110	10
At end			
Weight (g)	heaviest carp	1870	1205
	lightest carp	1170	510
Increase (g)	heaviest carp	1705	1165
	lightest carp	1060	500
Increase (%)	heaviest carp	1033.3	2912.5
	lightest carp	963.6	5000.0

SHOOT CARP
EXPERIMENT 69/5

Experiment 69/5 is a further attempt to investigate shooting in fish. Twenty full sibling carp were divided into two groups. Each group was allocated to two tanks with five carp each. The average individual weights were the same in all four tanks. However in one group all fish weighed almost the same at the start of the experiment, while in the other group the individual weights varied widely. All fish were individually marked and are listed in the data and results table, in the same order throughout. The experiment produced the following results:

1. The carp of equal weight at the start of the experiment grew apart in weight considerably during the roughly 4 months of the experiment. The difference between extreme weights in tank 69/5−2 reached 220 g. The distribution within the group is illustrated in Fig. 68.

2. There was no change in the weight sequence at the end of the experiment within the carp group which had varying individual weights at the beginning of the experiment. Although the lightest fish in 69/5−4 achieved nearly the highest percentage increase in weight of all fish in the experiment it remained at the bottom of the range.

3. At the end of the experiment the average weight of all carp with the same weight at the start was higher than that achieved by the group with differing weights at the start.

If it is assumed that behavioural factors cause weight classing then this experiment shows that a new hierarchy is established in a tank within a period of some months at least and that within the same time span existing hierarchies do not break down.

MATERIAL AND METHOD

Duration	124 days
Fish	carp from mating 8036
Method	69/5—1 and 2: 5 carp each of same weight at start
	69/5—3 and 4: 5 carp each of different weight at start
Feed	trout diet A
System	circuit 504
Tank	40 l, plastic
Water flow rate	2.5 l/min

DATA AND RESULTS

	Fish No.	69/5—1 Carp with same weight at start	69/5—2 Carp with same weight at start	69/5—3 Carp with different weights at start	69/5—4 Carp with different weights at start
At start					
Individual					
weights (g)	1	190	190	300	305
	2	185	185	255	255
	3	190	190	140	135
	4	185	185	145	140
	5	185	185	95	95
$\bar{x}$ weight (g)		187.0	187.0	187.0	186.0
At end					
Individual					
weights (g)	1	595	630	805	820
	2	565	530	655	635
	3	650	540	400	580
	4	630	750	400	530
	5	dead	560	250	365
$\bar{x}$ weights (g)		610.0	602.0	502.0	586.0
Increase (g)	1	405	440	505	515
	2	380	345	400	380
	3	460	350	260	445
	4	445	565	255	390
	5	dead	375	155	270
Increase (%)	1	213.2	231.6	168.3	168.9
	2	205.4	186.5	156.9	149.0
	3	242.1	184.2	185.7	329.6
	4	240.5	305.4	175.9	278.6
	5	dead	202.7	163.2	284.2

TECHNIQUES FOR BREEDING PROGRAMMES

MARKING

It is necessary to identify individual fish in work on breeding. It allows fish with differing genetic backgrounds to be reared under the same environmental conditions. At Ahrensburg all carp were marked well before the time they were first used for breeding or multiplication.

A very simple piece of equipment was used. It consisted of a near-circular loop of strong wire which was heated from an accumulator such as a car battery. It had a wooden handle for easy use, and was operated by a switch. Depending on how the wire loop is applied the fish can be branded with circles and bars. A combination of these basic figures on different parts of the fish's body made it possible to identify individual fish by numbers up to 1000 (Fig. 69).

The carp being numbered were first anaesthetized with Sandoz MS 222 and the hot wire was briefly but firmly applied to its body. Provided that contact was brief and only the epidermis was affected the fish suffered no harm. The superficial lesion heals quickly while the scar remains visible for several years.

The potential use of techniques described in this book for breeding programmes, especially fertilisation *in vitro*, can be illustrated by the following examples.

Using the *in vitro* insemination method described under breeding and breeding technology (page 37) the production of full sibs presents no problems. This cannot be achieved in ponds where several males or cock fish are introduced to each female or hen fish in order to induce successful courtship display and mating (Wunder, 1966). It is equally easy to obtain half-sib groups where one female is paired with several males, or one male with several females at the same time (see Meske, 1968b) (Fig. 70).

Another possibility is the production of crosses between species by stripping mature fish, sometimes after treatment with hormones. The methods for obtaining eggs and milt and the rearing of the resulting fry have been described previously in Chapter 4.

The eggs of a female carp were divided between two bowls. The eggs in one bowl were fertilized with the milt of a male goldfish (*Carassius auratus auratus*) and those in the other with carp sperm. Figure 71 shows the carp/goldfish cross and Fig. 72 the pure carp cross. Both fish had the same mother. It was possible to repeat these experiments in a number of cases. Of particular interest are the crosses between a female Prussian carp (*Carassius auratus gibelio*) and a male tench (*Tinca tinca*), male roach (*Scardinius erythrophtalmus*), male carp (*Cyprinus carpio*) and male goldfish (*Carassius auratus auratus*) which Cellarius (1973) produced. Gynogenesis has been attributed to the Prussian carp but it is reported by several authors (e.g. Ladiges and D. Vogt, 1965) that it uses the sperm of other species as a mechanical activator for egg development, the female Prussian carp mixing with spawning fish or related species. In fact in all four cases of crossing referred to above, all progeny were phenotypically pure Prussian carp.

Using currently available breeding techniques inbred lines can be developed. The progeny of a single pair of fish can be reared separately until it is possible to brand them, and in the course of several generations it is then possible to establish almost pure but distinct lines or strains. In the process inbreeding depression and deformations will inevitably be met.

Over the years the experiments produced some fish with abnormal gonads which included transexual and bisexual organs (Gupta and Meske, 1976). Figure 73 shows the gonads of three

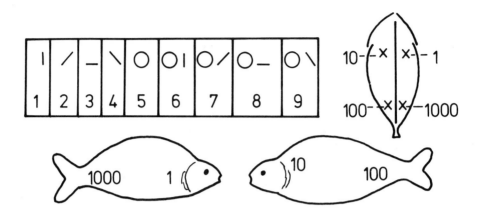

Fig. 69: Branding breeding stock. *Top left*: Number code
(1–9). *Top right*: Plan view of location for digital coding.
Bottom: As top right but seen from sides.

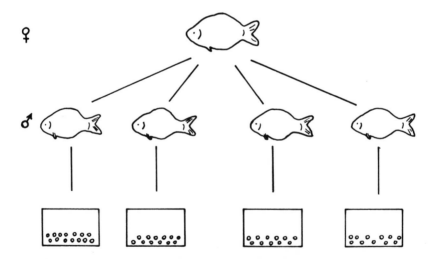

Fig. 70: Example for obtaining half-sib groups. The eggs of one female
are inseminated with the milt of several males *in vitro*. The performance of the progeny gives an indication of the parent's genetic
potential.

Figs. 71 and 72: Fish frow two *in vitro* produced half-sib groups. Fig. 71: Carp/goldfish cross (♀: carp no. 60. ♂: goldfish no. 94). Fig. 72: Pure carp (♀: carp no. 60. ♂: carp no. 40).

carp with distinctive female and male tissue sections. In 1971 Kossmann found an autogamous hermaphrodite which yielded fertile eggs and sperm. Among the progeny of this fish, hermaphrodites occurred only sporadically, some of which distinguished themselves by displaying very high productivity (Hilge and Conrad, 1975).

If it were possible to establish inbred lines from autogamous hermaphrodites it would be possible to achieve genetic improvement very quickly, as is the case for instance with self-fertilizing plants. Figure 74 illustrates one bizarre effect of inbreeding observed by Kossmann (1970): the tail-less carp which showed no lack of vitality.

BREEDING FOR IMPROVED PERFORMANCE

The examples discussed below are only an indication of the many possibilities in the field of genetic work on the carp kept in warm water. Because of the limited housing resources at Ahrensburg the research on fish breeding was restricted to those aspects which are described in this book. The decisive factor in work of this nature is the ability to rear the progeny in a controlled and preferably standardized environment. The performance of the progeny can then be ascertained and compared.

To improve the performance of carp through a breeding programme — selection and testing — the following need to be considered:

1. growth rate;
2. survival rate;
3. feed conversion ratio;
4. lean meat content;
5. length—height ratio.

There is probably some correlation between items 4 and 5, but this requires further investigation. The traditional subjective selection criteria no longer apply. Oil content is no longer an important selection parameter because its control can be achieved easier and more quickly by appropriate management and feeding than by selective breeding. In the case of selection for resistance to disease, Kirpichnikov *et al.* (1979) reported on abdominal dropsy in carp.

Up until now the methods of improving pond culture fish have relied mainly on observation of visual characteristics — phenotypes — and the selection of shooters (Probst, 1943; Wunder, 1956, 1960). Hence stock improvement in fish lags far behind that achieved in warm-blooded farm stock — cattle, pigs, poultry and rabbits. The application of current knowledge of genetics was introduced into carp breeding by Moav and Wohlfarth (1960). They were able to demonstrate that the performance of progeny of shooters regresses and that the progeny of non-shooters may outstrip those of the shooter. The use of scientific genetics also confirmed that selection and multiplication of the bigger individuals of a population could not produce the expected success, as apparently the heritability of growth in carp is low (see, e.g. Moav, 1979). Instead progeny testing may be more satisfactory where the breeding stock is selected on the basis of the performance of their progeny. Progeny testing makes standard conditions obligatory.

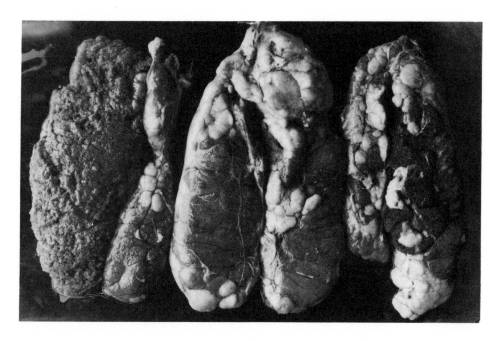

Fig. 73: Gonads of three bi-sexual carp. The darker ovaries are interspersed with the lighter testicular tissue.

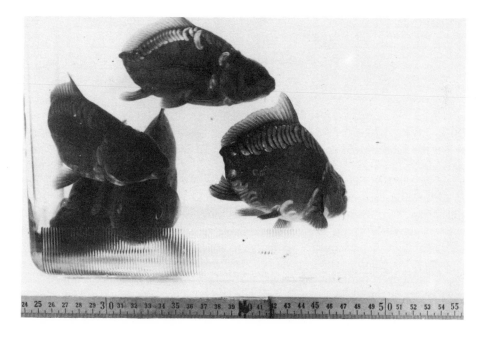

Fig. 74: Tail-less carp. These carp appeared during work on inbreeding.

PROGENY TESTING AT AHRENSBURG

Progeny testing of carp at Ahrensburg has been carried out for some years. Testing is conducted under the most stringent standard and constant environmental conditions of the warm water installation using the following programme:

1. After hatching, 200 progeny of one pair of parent fish are taken at random and placed in a 40 l through-flow aquarium.

2. At age 4 weeks 100 of the 200 fish are selected at random for progeny testing. The fish are individually weighed.

3. At age 8 weeks the 100 fish are moved to an 80 l aquarium which was part of the closed-circuit system. The 100 fish are again individually weighed.

4. At 12 weeks the fish are weighed again individually and their overall length measured. The fish are divided into two groups of equal numbers and weight and placed in two 80 l aquaria.

5. At 16 weeks the test is concluded – each fish is weighed and measured.

Five fish are selected and placed in a 40 l aquarium. These five fish continue to be weighed every 4 weeks. When necessary – i.e. because of size – the stocking density of the aquarium is adjusted to ensure the fish are provided with an optimal environment.

The feeding of fry in progeny tests follows this programme: Fry are first given *Artemia salina* between 07.30 and 16.30 hours daily. 14 days after hatching they are gradually weaned on to a concentrate consisting of 50% salmon fry feed A and 50% trout fry feed B. The analyses of these feeds were:

	Protein (%)	Oil (%)	Fibre (%)
Salmon A	49	17	1.5
Trout B	52	8	2.0

The change-over is achieved by feeding *Artemia* and concentrate alternatively. After this the fry are fed the salmon-trout concentrate mix only until they are switched to 100% trout fry concentrate B at 12 weeks.

Because of the large number of progeny produced by table fish they should be an ideal subject for breeding programmes. However the complex environment of ponds or rivers is a major handicap. Even when efficient marking was introduced (Moav *et al.*, 1960) constructive breeding was prevented by the many uncontrollable variables. The difficulties of selection and the implementation of performance testing were appreciated, for instance, by Bakos (1967) in his guide on the improvement of pond fish. The breeding of carp in warm water offers many advantages and opportunities to improvers, particularly the development of inbred lines and strains which are produced for a specific quality or goal. On crossing they should display hybrid vigour which may be superior in effect to that found in poultry. Results observed to-date are given by Bakos (1979).

Purdom (1969, 1972) pioneered new ground in fish genetics when he produced polyploid flatfish by subjecting the fertilized eggs to cold-shock treatment. In eastern Europe work on the gynogenesis of carp and other cyprinids has been successfully carried out (Golovinskaya, 1968; Nagy *et al.*, 1978).

BREEDING OUT INTRAMUSCULAR BONES

Based on his experimental work in plant breeding, von Sengbusch (1963) conceived a plan to breed out the intramuscular bones in certain fish species. According to the concept of parallel variations certain characteristics which normally occur in related species, while being absent in others, can be expected to occur very occasionally in the latter, and this von Sengbusch exploited. The classic example is his breeding of the sweet lupin after the discovery of individuals devoid of alkaloids. Morphological modifications were achieved too, through applying this concept, an example being a lupin which does not shed its seeds.

This approach requires a very large number of individuals to be examined until a specimen with the required characteristics is found. In the case of the lupin the frequency of alkaloid and non-alkaloid specimen occurring lies somewhere between 1 in 20 000 and one in a million. For the non-shedding lupin pod the ratio is 1:10 million. Von Sengbusch (1967a) also describes anthocyanin-free asparagus, parthenocarpic tomatoes, nicotine-free tobacco, mushrooms without lamellae and monozotic spinach and asparagus.

The carp belongs to that group of fresh water fish considered to have an average number of bones. Lieder (1961) counted an average of 99 intramuscular bones in the carp, 21 in the ruffe (*Acerina cernua*) and 143 in the asp (*Aspius aspius*). All other 14 fish species examined by him fell within these values. The ruffe has a deep back, whereas the asp is slim. It is therefore difficult to ascribe a body-stabilizing function to intramuscular bones. The many intramuscular bones of the carp are ossified and connect tissue outside the skeletal system (Fig. 75). As other fish — such as the Atlantic cod (*Gadus morrhua*), for instance — do not have this type of bone (Fig. 76) the existence of carp mutants without intramuscular bones could be expected.

The research work required equipment with the ability and sensitivity to detect the parameters being investigated. These criteria were met by an X-ray television camera and monitor which picked up and displayed fine bones clearly. Carp of 10 g upwards could be examined, with the added advantage that live anaesthetized fish can be examined very quickly compared to the time-consuming X-ray photography method. Figure 77 shows the equipment which made examination of several thousand fish a day possible, provided the supply logistics were well organized. The equipment was used for a number of years.

Altogether some 13 000 carp were examined. Although this number is small compared to the incidences in plants referred to above a considerable spread of the number of bones was established (von Sengbusch and Meske, 1967).

The number of bone points or ends in 704 6-month-old carp of a pond population ranged from 70 to 134, with a mean of 100 bone points. The intramuscular bones branch on ageing. One hundred 6-month-old carp, and 100 18-month-old carp were examined for the number of bone bases and bone points. It was found that the 6-month-old carp had an average of 100 points and the 18-month-old carp averaged 129 points.

Theoretically there are two ways of breeding carp without intramuscular bones:

1. The number of bones can be reduced by systematic selection and inbreeding until a low-

Fig. 75: X-ray photograph (detail) of a carp showing vertebrae, spinal transverse processes and across these the intramuscular bones (Sengbusch, 1963).

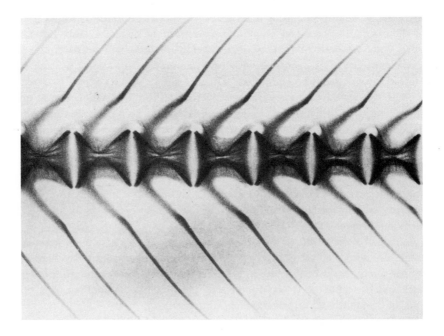

Fig. 76: X-ray photograph of an Atlantic cod (*Gadus morrhua*) corresponding to the above. This species has no intramuscular bones (Sengbusch, 1963).

bone carp is produced from which it may eventually be possible to breed a bone-less carp.

This would involve numerous generations and success would be a long way off. According to Kaendler (1971) pond-cultured mirror carp already have fewer bones than unimproved wild strains. Kossmann (1972) found statistically significant differences in the number of intromuscular bones in some groups of full-sib carp. In a similar investigation on over 1000 carp Moav *et al.* (1975) however found only distinctly small variations in bone numbers. They expressed the view that selective breeding for this characteristic was not worth while.*

2. If it is assumed that only one or a few genes are responsible for the existence of intramuscular bones, then occurrence of extreme individuals devoid of these bones can be expected (von Sengbusch, 1967a). The distribution of intramuscular bones found during the work at Ahrensburg has similarity with a Gaussian error curve and this would indicate that mutants would occur in a new peak completely outside this curve. Any hope of finding a mutation would presumably require the examination of a very large number of fish, say several hundred thousand individuals.

In the majority of successes attained by von Sengbusch in plant selection, extreme forms with the desirable properties occurred without intermediate transitional forms (e.g. non-shedding lupin pods, mushrooms without lamellae).

Hence the second of the two options would appear to be more likely to yield satisfactory results. During recent years, work at Ahrensburg on developing new techniques and improving management, feeding and multiplication systems has been given preference to resource-intensive genetic work on the qualitative improvement of any specific characteristic. In particular the search for carp without intramuscular bones had low priority because of the high labour and financial input required. Instead work has concentrated on table fish with few or no intramuscular bones such as the European wels (*Siluris glanis*), tilapias spp. and the pirarucu (*Arapaima gigas*), among others. This work is described in Chapter 6.

---

* Editor's note: Vogt (1979, 1980, 1982) suggests an analogy with farm animal husbandry that the incidence of secondary bones or bone points in carp can be eliminated by management which could produce a marketable portion size fish in 12 months and before forking occurs.

Fig. 77: Television monitor used for counting
intramuscular bones. Below it the controls.

Fig. 78: Eel cage battery. Faeces and waste feed are removed from the
tank through the suction pipes on the left.

6. EXPERIMENTS ON THE MANAGEMENT, FEEDING AND PROPAGATION OF OTHER WARM WATER CULTURED TABLE FISH

EELS (*Anguilla anguilla*)

MANAGEMENT

The following experiments were undertaken to clarify the development of the individual eel. The eels were raised from the glass eel stage onwards. Large eels mostly were not used because they are adapted to a different environment and present additional problems.

The eels' migratory instinct causes major difficulties. The commonly used aquaria allow the fish to escape after being caught. Hence steel frame cages covered with fine wire mesh and with a closely fitting lid were made (Meske, 1969). Several of these cages were suspended in each aquarium (Fig. 78) and proved satisfactory for the first feeding trials. In later experiments it was found possible to keep the eels in an ordinary aquarium with a secure glass top and the inlet and outlet pipes covered with wire mesh.

Behavioural problems are major factors during the rearing of eels in aquaria and tanks. The fish become very aggressive and biting is common. Cannibalism is a frequent occurrence among elvers. This was the cause of significant losses in tanks.

It is therefore useful to provide the eels with some cover, which in the case of the glass eels was plastic tubing drilled with holes. The eels tend to move in after only a few hours, their heads protruding through the holes and they then remain relatively peaceful. For larger eels a plastic pipe laid on the bottom of the tank is sufficient to give them protection from each other.

For certain experiments with single eels it was necessary to cage them individually. For this purpose a radial tank (Fig. 79) was designed, fitted with a revolving transparent lid. The radial tanks save space and materials.

THE INFLUENCE OF LIGHT ON EELS
EXPERIMENT 82/2

Intensively kept eels are raised in normal daylight conditions and fed during the daytime. This experiment was carried out to ascertain whether this regime provides the best conditions for eel growth. The results show the differences in growth due to four lighting and feeding regimes.

Figure 80 illustrates the growth range achieved. Eels under continuous light over the period of the experiment grew least. Eels on a 12 hour day/night cycle fed during the light hours grew to an average weight of 201 g. The best growth was achieved by eels on the 12 hour day/night cycle fed in the dark while shining a weak torch for 1 minute. They ate more of the floating pellets and converted more efficiently than the eels in the other three groups in the experiment. At the end the weight of these eels averaged 231 g.

Over the 18 weeks of the experiment about 30% of the eels died from various causes, such as biting and tumours. The fish died singly, predominantly through biting. Their growth up to their death is included in the data on weight increases.

No firm conclusions can be drawn from this experiment. The eels were not uniform and salmon concentrate is not the most suitable feed. The result was weak growth and only moderate feed conversion efficiency. Nevertheless it is reasonable to assume that the influence of light on the nocturnally active intensively kept eel merits attention.

MATERIAL AND METHOD

Duration	239 days
Fish	eels from an Ahrensburg brackish water circuit gradually adapted to fresh water
Programme	82/2−1 + 2 Room 1: 24 hours light/day
	82/2−3 + 4 Room 2: 12 hours light/day
	from 06.00 to 18.00 hours
	82/2−5 + 6 Room 3: 12 hours light/day
	from 18.00 to 06.00 hours
	82/2−7 + 8 Room 4: no light
	45 eels per tank
Lighting	two 40 W Osram L40 W/15 strips per room
Water temperature	23°C
Water flow rate	13 l/min
Tank	250 l (1 m x 1 m x 25 cm depth)
Weighing	every 4 weeks
Feeding and feed	twice daily, manually, at 07.30 and 16.30 hours at rate of 1% of live-weight per day. Salmon concentrate analysis: protein 46%, oil 14%, fibre: 2.3%, ash 9%

Fig. 79: Radial tank for housing eels individually. Water flows into the compartments through holes in the central pipe, and leaves it through holes on the periphery into a radial gutter which leads to the drain. The radial tank is covered by a transparent rotating lid with a hinged door the size of one compartment.

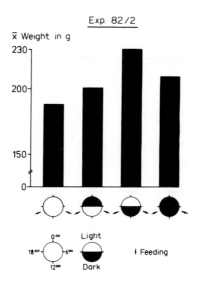

Fig. 80: The influence of light on eels
(Experiment 82/2).

DATA AND RESULTS

| Experiment code No. | Duration of lighting period | Fish numbers and weights | | | | | | Weight increase | | | FCR |
| | | at start | | | at end | | | | | Feed fed | |
		No.	Total weight (g)	Mean weight (g)	No.	Total weight (g)	Mean weight (g)	Total (g)	(%)	(g)	
1 + 2	24 hours/day	90	13 198	146.64	64	12 039	188.11	2301	17.44	18 047	7.84
3 + 4	12 hours/day 06.00–18.00	90	13 190	146.56	66	13 320	201.82	3352	25.41	18 579	5.54
5 + 6	12 hours/day 18.00–06.00	90	13 186	146.51	59	13 644	231.25	4580	34.73	18 660	4.07
7 + 8	no light*	90	13 179	146.43	63	13 235	210.08	3928	29.81	18 710	4.76

* Torch light for 1 minute during each feeding.

FEEDING

Feeding glass eels with minced fish proved the best method. Both sea fish and fresh water fish were used. The fast-growing eels were successfully fed with the aid of a perforated pipe suspended vertically in the aquarium. The feed was pushed down the pipe with a plunger as required. There is very little wastage with this system (Fig. 81). Feeding dry concentrate is more difficult with eels than with carp. Granular floating feed is mostly wasted. Finely ground dry feed mixed with water, or better still with cod liver oil, can be worked into a paste and stuck on to the aquarium's side (Fig. 82). This is eaten with avidity. A floating-mesh frame offers many advantages. The fish obtain their feed throught the mesh from above the water line. This method is used in Japan.

At Ahrensburg the most effective feed for good growth in eels proved to be wet fish. Larger eels were fed entire minced carp, offal included. The mince was made up into portions and kept in a deep freeze. It was fed unthawed straight into the warm water aquaria. This method proved very successful. In contrast, young eels with an average initial weight of 0.88 g fed exclusively on dry concentrate achieved only minimal growth and high losses. This was demonstrated in an experiment involving 1200 young eels. After 5 months the controls fed on carp meat had a loss rate of 27.5% and the experimental group fed a Japanese dry eel feed had lost 84.3% (Meske, 1973a). In the last years worms (*Tubifex* sp.) and new types of dry food were used successfully (see below).

It is known that young eels ascending rivers generally grow quite slowly. Meyer-Waarden (1965) quotes Ehrenbaum's growth rate table. Several other workers, including Deelder (1970), have confirmed these data, claiming that ascending elvers attain a weight of 1 g after 0–1½ years in fresh water, 3 g at 2–2½ years, etc.* However, Mueller (1967) reports a pond experiment in which the biggest eel weighed 20 g at the end of the first summer. This would indicate that Ehrenbaum's table is open to doubt.

Eels raised at Ahrensburg in warm water at 23°C showed an even better growth in the same time span. Starting as glass eels these fish reached individual weights of 124 g, 110 g, 92 g, 83 g, 81 g, etc. after 7 months (Meske, 1968d). The most vigorous of these eels put on 100 g in 8 weeks, growing from 24 g to 124 g. Figure 83 shows the growth of this fast-growing fish compared with the maximum growth observed in ponds. Figure 84 represents the individual weights of the 23 heaviest eels of an experimental group of 223 eels after 7 months of warm water culture. After 13 months the heaviest eel in the group weighed 275 g. In terms of Ehrenbaum's table this is equivalent to a 9–9½-year-old well-developed female in natural waters.

Figure 85 emphasizes the difference between observed maximum weight after 13 months of warm water culture (lower picture), the pond fish size as reported by Mueller (1967) (centre) and a 2 g eel (top) corresponding approximately to Ehrenbaum's table for growth in natural waters. The extraordinary rapid weight increase found in warm water culture is also noticeable during the second year. The biggest of the glass eels put into aquaria in February 1970 weighed 583 g on 15 October 1971. In contrast Koops (1965) reported feed trials involving 7500 glass eels in ponds which reached maximum individual weights of only 57 g at the end of the second summer.

Several batches of glass eels have been reared at Ahrensburg. The most startling observation made is the wide weight range of the fish. After 1½ years of warm water culturing the spread of weights of 225 eels ranged from 2 g to 520 g with an average weight of 50.6 g (Meske, 1969c).

It is not yet clear what causes this spread of weights. External influences must be discounted because standard conditions prevailed in the aquaria. It seems that in tanks with high stock density the eels develop a dominance hierarchy or pecking order, and that the fast growers among these fish soon exploit their weight advantage further. Biting among the fish was regularly observed, which turned into cannibalism when feeding was inadequate. It was thought possible that the initial size difference in the glass eels may be due at least in part to the differential growth rates of the sexes. Males are also said to develop more slowly. Sex determination is perhaps not yet fixed at the larval or leptocephalic stage. It has been reported that eels at river mouths consist almost entirely of males, while catches from tributaries of large rivers are nearly 100% female (Deelder, 1970).

Males predominate in Japanese pond culture (Koops, 1967) although they are economically of less interest because of their lower growth performance. The development into male and female fish could possibly be due to environmental factors such as salinity or feed composition. It may be possible to influence the sex of small eels by feeding small amounts of hormones befor sexual differentiation is completed, as has been achieved by Purdom (1969) with trout.

However it seems that neither the variation in weight nor the growth rate is linked with the sex of the eel. Experiment 69/44 examined the growth rate of two age groups, both with the same average liveweights at the start of the experiment.

---

* Editor's Note: The growth rate varies with latitude. River Severn eels take 7–15 years to reach maturity while Scottish eels may take up to 30 years.

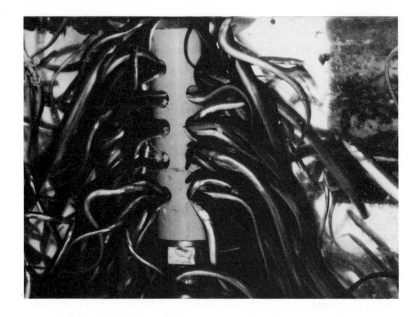

Fig. 81: Feeding eels. Feeder consists of a pipe drilled with holes from which eels feed. Additional feed is pushed into the top of the pipe as required.

Fig. 82: Feeding eels with dry concentrate. The dry feed is mixed with cod liver oil and the paste is then stuck to the aquarium's wall.

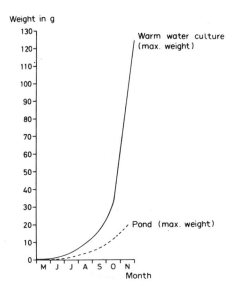

Fig. 83: Maximum observed growth of eels. During their first
year in fresh water (stocked as glass eels) in warm water (solid
line) (Meske, 1968d) and in ponds (Mueller, 1967).

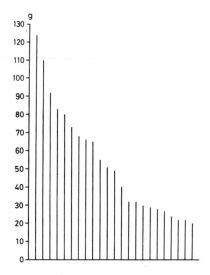

Fig. 84: Individual weights of the 23 heaviest eels in a group
after 7-monthly warm water culture. Average weight at start:
0.4 g.

FEEDING GLASS EELS
EXPERIMENT 80/2

To gain a better understanding of glass eel feeding, four feeds were tested:

brine shrimp (*Artemia salina*) nauplii;
minced whole carp (wet);
tubifex worms;
eel concentrate.

Each treatment involved 200 glass eels in two groups of 100 each contained in a 20 l glass aquarium. The fish were kept in mains flow-through water at 23°C.

The first part of the experiment ran for 2 months. The results (Part I) show that only the group fed tubifex worms increased their weight (+ 156.64%). The fish in the other three groups lost weight. The tubifex group also suffered, at 4%, the least fatalities. The group on eel concentrate had the biggest fatality rate (62%), mainly because of cannibalism.

At the end of the first part of the experiment the fish in one of the two aquaria in each group were switched to the eel concentrate. The other half of each group continued to receive the original treatment.

The overall results (Part II) cover the 114 days duration of the experiment. Again the fish on tubifex showed the best weight gains at 422.34%. The next-best result was achieved by the tubifex/eel concentrate sub-group with 189.9%. Third came the eel concentrate fish with only 77.83%. The eels fed brine shrimp nauplii and minced carp lost weight.

Fatalities were 18% in the tubifex only group, 62% in the tubifex/concentrate sub-group and 77% in the concentrate only group.

The glass eels fed tubifex achieved by far the best weight gains compared with the other feed regimes in the experiment. The fish on eel concentrate had a high incidence of cannibalism and made only moderate growth.

MATERIAL AND METHOD

Duration	114 days
Fish	glass eels of intake 2.80
Programme	

80/2–1	commercial eel concentrate	
80/2–2	brine shrimp nauplii	for first 2 months,
80/2–3	minced whole carp	then changed on to
80/2–4	tubifex	eel concentrate
80/2–5	commercial eel concentrate	
80/2–6	brine shrimp nauplii	
80/2–7	minced whole carp	
80/2–8	tubifex	

each tank contained 100 eels

eel concentrate analysis: protein 58%, oil 17%, fibre 0.7%, ash 5.7%

Tank	30 l aquaria (approx. 10 cm depth of water) with PVC pipes for cover
Feeding	first 20 days: 2.5% of liveweight, fed twice daily
	from 21st day onward: 3.75% of liveweight, fed three times per day
Water temperature	23°C
Water flow rate	2 l/min approximately, mains water, flow-through
Weighing	every 4 weeks, alternatively total and individual weight

DATA AND RESULTS

Experiment code No.	Feed	No. and weight of fish						Fatalities	Total weight gain/loss	
		at start			after 60 days					
		No.	Total weight (g)	Mean weight (g)	No.	Total weight (g)	Mean weight (g)	(%)	(g)	(%)
Part I (first 2 months)										
1 + 5	Eel conc.	200	61.66	0.31	76	38.70	0.51	62	− 22.96	− 37
2 + 6	Brine shrimp	200	57.03	0.29	160	38.43	0.24	20	− 18.60	− 32
3 + 7	Minced fish	200	56.73	0.28	111	53.22	0.48	44.5	− 3.51	− 6
4 + 8	Tubifex	200	59.73	0.30	182	153.29	0.84	8	93.56	157
Part II (overall)										
1 + 5	Eel conc.	200	61.66	0.31	46	109.65	2.38	77	47.99	78
2	Brine shrimp + eel conc.	100	30.00	0.30	26	35.73	1.37	74	5.73	19
3	Minced fish + eel conc.	100	28.66	0.29	25	41.25	1.66	75	12.88	45
4	Tubifex + eel conc.	100	28.67	0.29	38	83.12	2.19	62	54.45	190
6	Brine shrimp	100	27.03	0.27	51	25.25	0.50	49	− 1.78	− 7
7	Minced fish	100	28.07	0.28	29	27.56	0.95	71	− 0.51	− 2
8	Tubifex	100	31.06	0.31	82	162.24	1.98	18	131.18	422

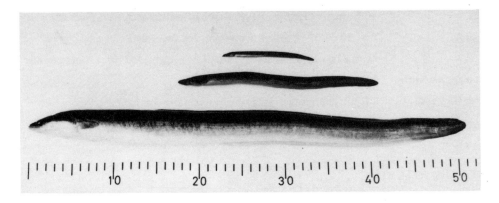

Fig. 85: Three eels after 13 months in fresh water. Eel at the top represents average growth in the natural environment and weighs 2 g. The eel in the centre at 20 g shows maximum growth achieved in a pond during the same period. The eel at the bottom is from a warm water culture and weighs 275 g.

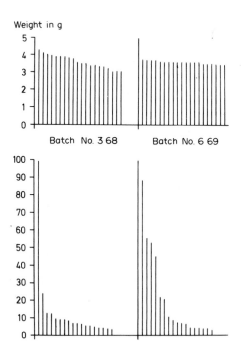

Fig. 86: Growth rate of eels of different age groups (Experiment 69/44). Weights of eels from two age groups at start (top) and at end (bottom) of experiment lasting 564 days. *Left column*: glass eels from batch No 3 68 of 1968. *Right column*: glass eels from batch No 6 69 of 1969.

THE GROWTH RATE OF EELS OF DIFFERENT AGE GROUPS
EXPERIMENT 69/44

Experiment 69/44 produced somewhat unexpected results. The fish of batch 3 68 were caught as elvers in 1968, those of batch 6 69 in 1969. Both were warm water cultured in separate aquaria. Although the eels belonged to different age groups they were the same average weight at the start of the experiment. During the 18 months of the experiment these eels developed at a rate inverse to their age. The 1 year younger fish showed a statistically highly significantly greater weight increase than the older fish (Fig. 86). Furthermore, at the end of the experiment the variation in individual weights was narrower among the younger fish than among those 1 year older, which were dominated by a fish of 98 g (see table below). The results of the experiment may indicate that at the start, the younger eels were all shooters which reached starting weight after only 6 months. The reason for this burst of growth in this group is not clear. Theoretically it is possible that these eels were mainly females but there may be other causes, such as the different origins of the catches. This is indicated by the results of other experiments not described here.

TABLE 12

Individual weights (g) of eels of two age groups (Fig. 86)

69/44–1: Batch 3 68		69/44–2: Batch 6 69	
At start of experiment	At end of experiment	At start of experiment	At end of experiment
4.30	98.0	3.75	88.4
4.10	24.1	3.70	55.8
4.00	12.6	3.70	53.9
3.95	12.5	3.70	45.3
3.90	9.5	3.65	22.4
3.90	9.2	3.65	21.1
3.90	9.1	3.60	11.0
3.85	8.8	3.60	9.4
3.80	7.2	3.60	7.9
3.55	7.1	3.60	7.2
3.50	6.7	3.60	6.7
3.50	5.9	3.60	4.7
3.40	5.7	3.60	4.6
3.40	5.3	3.60	4.5
3.35	4.5	3.50	4.3
3.35	4.5	3.50	4.2
3.20	4.3	3.50	3.1
3.05	3.8	3.50	
3.05		3.45	
3.05		3.45	

It would appear desirable to test the effect of sex differentiation of eels on growth in aquaria. If sex is of overriding importance and can be influenced by environmental factors or by feeding, the production of female eels only would be of economic interest. For this reason experiments to determine sex through the feeding of hormones to glass eels are of interest.

MATERIAL AND METHOD

Duration	200 days
Fish	eels of batch 3 68 and 6 69
Method	69/44—1: 20 eels from batch 3 68
	69/44—2: 20 eels from batch 6 69
Feed	minced frozen carp
Feed method	twice daily (7.30 and 17.00 hours)
Tank	30 l, glass
Water flow rate	4 l/min
Weighing	every 28 days

DATA AND RESULTS

	69/44—1 Batch 3 68	69/44—2 Batch 6 69
At start		
Number	20	20
Weight (g)	72.1	71.8
$\bar{x}$ Weight (g)	3.6	3.6
At end		
Number	18	17
Weight (g)	238.8	354.5
$\bar{x}$ Weight (g)	13.3	20.9
Weight increase (g)	173.8	293.5
Weight increase (%)	267.4	481.2
$\bar{x}$ Weight increase (g)	9.7	17.3

ANALYSIS OF VARIANCE Exp. 69/44

	FG	SQ	MQ	F
Total	34	18 447.05		
Age groups	1	11 063.62	11 063.62	49.45**
Residual	33	7 383.43	223.74	

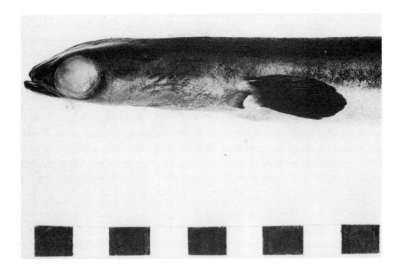

Fig. 87: Mature male eel after receiving hormone treatment. Eyes and pectoral fins
are greatly enlarged.

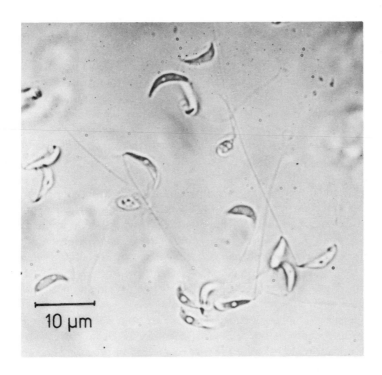

Fig. 88: Free-swimming eel sperm.

PROPAGATION

Little is known about the development of the European eel during its early years of life. There-fore attempts at propagating eels in aquaria were and are being undertaken. The first essential is to achieve sexual maturity, and in the case of male eels this was successful (Meske and Cellarius, 1972, 1973).

Several groups of glass eels were put into warm water aquaria in 1968. In December 1971 they were placed in sea water tanks with through-flow recirculation at a water temperature of 22°C and a salinity of 35‰. Almost immediately the fish ceased feeding completely. From April 1972 onwards the fish of one experimental group were each given intramuscular injections at roughly 2-week intervals of 0.2 ml Solcosplen extract and 50 rabbit-units of Synahorin, a Japanese preparation combining gonadotrophic hormones from the frontal lobes of the pituitary and the placenta, in a ratio of 1:9.

Solcosplen is a protein-free extract of fresh calves' spleen, the positive effect of which on the development of the gonads of rodents has been reported by Grigoriadis et al. (1969) and Goslar et al. (1969). According to these authors the spleen extract has a considerable stimu-latory effect on, for example, infant guinea pig testes. The Leydig cells were greatly enlarged, the majority of the tubuli opened and lactate dehydrogenase and 3-ol-steroid-dehydrogenase activity was elevated. The authors suggested that the active substance in the spleen extract has a stimulating effect upon the gonadotrophin production of the hypothalamus.

The treated eels soon displayed an enlargement of the eyes (Fig. 87) similar to a description by Stramke (1972) of some sea-caught eels. All males of this treated group emitted sperm on 3.7.72 when slight pressure was applied. Under the microscope this sperm was highly active. The weights of the fish were 155 g, 145 g and 125 g. The free-swimming sperm have a slightly bent oval and pointed shape (Fig. 88). They are about 9 μm long and have flagella about 30 μm long. Under the electron microscope (Meske, 1973b) this sperm of the European eel (Fig. 89) is very similar in structure to that of A. japonica, which was later examined under the electron microscope by Colak and Yamamoto (1974). One eel which died 7 days after emitting sperm had well-developed testes which represented 3.6% of its liveweight. Figure 90 is a view in situ of this mature male eel, whose testes run along its length in both sides of the abdominal cavity, most of which it occupies.

The treatment has been repeated successfully on a number of occasions. It even proved possible to obtain motile sperm from eels raised from elvers entirely in fresh water at 23°C. 3—4-year-old eels emitted sperm after being injected intramuscularly only three times with 0.2 ml of spleen extract and 50 RU Synohorin each. It was possible to obtain motile sperm from one eel six times in 4 months. Other workers have reported previously on successful artificial propagation, but these eels had not been raised from the glass eel stage to sexual maturity (Olivereau, 1961; Rontaine et al., 1964; Boetius and Boetius, 1967).

At Ahrensburg Hilge (1976) succeeded in stripping two female European eels after hormone treatment. The eggs were, however, unsatisfactory.

Nose (1971) reported stripping a female Japanese eel after hormone treatment. Yamamoto et al (1974a, b, c) describe hormone-induced maturity of male and female Japanese eels. Yamamoto and Yamauchi (1974) were the first to succeed in procuring eggs and fertile sperm with subsequent fertilization in the Japanese eel. They also described and photographed larval development (Yamamoto et al., 1975). This is a breakthrough which paves the way for planned

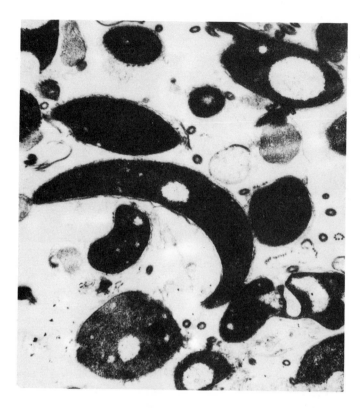

Fig. 89: Eel sperm under the electron microscope, enlarged 12 000 times. (*Source*: Schirren, University Clinics, Hamburg-Eppendorf.)

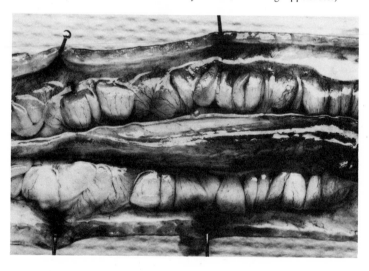

Fig. 90: Fully developed testes of an eel brought to sexual maturity experimentally.

Fig. 91: Grass carp (*Ctenopharyngodon idella*) raised on dry concentrate (Length 44 cm; weight 1022 g).

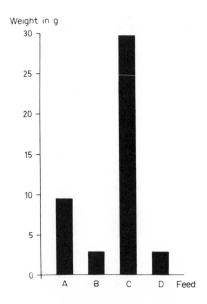

Fig. 92: Feeding grass carp (*Ctenopharyngodon idella*) with various feed-stuffs (Experiment 69/11). Average weight increase of grass carp after 98 days. A: Trout concentrate; B: mixture of 25% soya meal and 75% lucerne meal, pelleted; C: *Chironomus* larvae; D: duckweed (*Lemna minor*).

commercial eel production. The research group of Hokkaido University at Hakodate is continuing intensive work on the artificial propagation of the Japanese eel (Sugimoto *et al.*, 1976; Takahashi and Sugimoto, 1978), having reported on the rearing of experimentally obtained eel larvae up to the age of 14 days (Yamauchi *et al.*, 1976). It can now only be a question of time before artificial propagation of the eel on a commercial scale is a practical reality.

GRASS CARP (*Ctenopharyngodon idella*)

MANAGEMENT

Keeping grass carp (Fig. 91) in aquaria and plastic or asbestos tanks presents no difficulties. This fish is not aggressive and can be kept at high density like the carp. There are no problems in keeping it together with carp in the same tank. To attain good growth rates the grass carp should be kept at an even temperature: 23°C was used throughout the year in the experiments on its feeding and breeding described below.

FEEDING

To raise the grass carp intensively it is important to know their nutritional needs in terms of plant and animal food sources. Particular attention was paid to this aspect in the experiments.

FEEDING YOUNG GRASS CARP (C. idella) WITH FEEDS OF PLANT AND ANIMAL ORIGIN
EXPERIMENT 69/11

Young grass carp weighing 4.14 g each were fed the diets shown in Table 13, in aquaria. The experiment ran for 3 months and the results are shown in Fig. 92, which indicates that the best growth was achieved in this experiment when young grass carp were fed entirely on animal

TABLE 13

		Crude protein (%)	Oil (%)	Fibre (%)
A:	Trout crumbs containing fish meal	30.0	2.9	9.5
B:	Mixture of 75% lucerne (alfalfa) meal and			
	25% soya meal (as crumbs)	30.2	3.8	5.5*
C:	*Chironomus* (midge) larvae	60.0	2.0	—*
D:	*Lemna minor* (Duckweed)			

* *Editor's estimate.* Data not supplied by author.

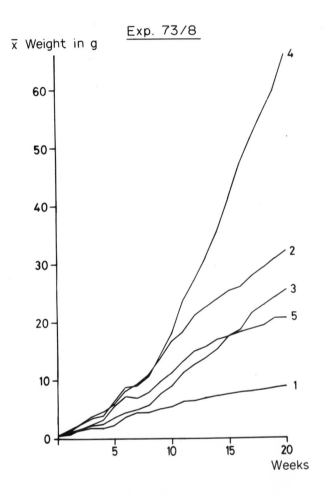

Fig. 93: Growth rate of young grass carp on various feed formulations (Experiment 73/8). Feed 1: trout concentrate, commercial; 2: SM3 mix containing no fish meal (see p. 00); 3: SM3 + lucerne meal; 4: SM3 + algae meal; 5: algae meal. For details see text.

protein (*viz*. Feed C: *Chironomus* (midge) larvae). At the end of the experiment the average weight increase of these fish was significantly better than that of any of the other groups in the experiment. Feeds of plant origin (Feed B: lucerne (alfalfa) and soya meal mixture and Feed D: *Lemna*) produced the least growth (Fig. 92). Grass carp, it seems, need animal protein for growth at least during their early life stages. Similar observations have been made by Huisman (1978).

In a further experiment 10 grass carp with an average weight of 246 g each at the start were fed exclusively with pelleted trout feed with an analysis of:

crude protein   40.1%
oil   4.3%
fibre   7.6%

The protein was mainly of animal origin (fish, meat, whale, liver and bloodmeal (see Meske, 1968c). The fish were kept at 23°C and grew on average to 729.4 g in 5 months. Two of these fish continued to be fed after the end of the experiment and attained weights of 1470 g and 1070 g respectively after a further 4 months. One of these fish fed trout feed containing fish meal weighed 10.5 kg after being kept in an aquarium for 5½ years. It then produced fertile eggs on hormone treatment as described elsewhere in this Chapter.

Experiments on grass carp showed that young grass carp grew best when raised on animal protein feed and that older fish grow well on high-protein dry concentrate. Since these fish had no access to any vegetation it must be assumed that young grass carp are predominantly carnivorous in the wild even when vegetation is available as food. This was observed by de Silva and Weerakoon during feeding experiments in 1981. In view of interesting results achieved with the carp it was considered to be worthwhile examining the effect of unicellular green algae of the genera *Scenedesmus* on growth when fed as a feed component.

In experiments 73/8 and 73/17 described below the effect of the green alga *Scenedesmus obliquus* as an additive to feed containing dairy products but no fish meal was investigated (Meske *et al.*, 1977). Roller-dried *Scenedesmus* algae, produced and supplied by the Institute for Research into Algae, at Dortmund, were used.

The growth of these fish during a period of about 4½ months is illustrated in Fig. 93. Particularly noticeable is the poor growth rate of the fish on trout Feed E in contrast to the very positive effect on growth of Feed H, which included 32% algae powder in a base feed containing no fish meal. Feed G (32% lucerne/alfalfa inclusion) and Feed I (algae exclusively) both resulted in only moderate growth.

When the appearance and health of the fish in the experiments were compared differences were clearly apparent, especially amongst the algae-fed fish. Figure 94 shows the fish on Feed H at the end of the experiment. They looked healthy and were of relatively even size. The fish fed entirely on algae (Feed I), on the other hand, displayed considerable deformities. Figure 95 shows their distinctly uneven growth, damaged fins and deformities of the spinal column. The grass carp fed on trout Feed E also grew unevenly, showing signs of malnutrition and some deformity (Meske and Pfeffer, 1978a). Each division of the scale at the bottom of Figs. 94 and 95 equals 1 cm, illustrating the difference in the size of the fish.

This experiment gave poor growth in fish fed entirely on algae, whilst a markedly beneficial effect was observed when algae were fed. Consequently further series of experiments were

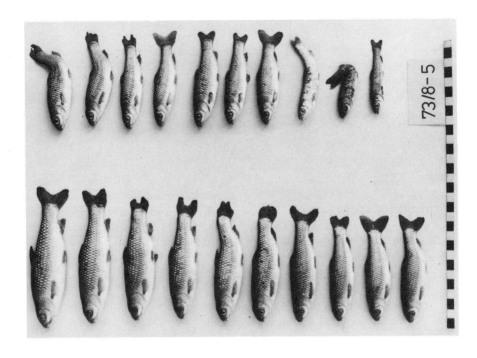

Fig. 95: Young grass carp fed exclusively on green algae (Experiment 73/8–5). See Feed 5 in Figure 93. Scale in centimetres.

Fig. 94: Young grass carp fed fish meal-free mixture containing 32% green algae (Experiment 73/8–4). See Feed 4 in Figure 93. Scale at bottom of photograph is 1 cm per division.

carried out. In these the algae content of the Feed F with no fish meal was systematically raised as shown below.

FEEDING GREEN ALGAE (Scenedesmus) AND MIXED FEEDS TO GRASS CARP (Ctenopharyngodon idella)
EXPERIMENT 73/8

MATERIAL AND METHOD

Duration	140 days
Fish	grass carp from mating 3044 hatched on 24 May 1973
Treatment:	73/8−1: 20 grass carp fed Feed E
	73/8−2: 20 grass carp fed Feed SM3 (meal)
	73/8−3: 20 grass carp fed mix of 68% SM3 and 32% lucerne
	73/8−4: 20 grass carp fed mix of 68% SM3 and 32% algae
	73/8−5: 20 grass carp fed algae only
Feed	Feed SM3 (without fish meal) and Feed E

Feed SM3:

Soya	391 g	Wheat	10%
Dried whey	509 g	Oats	7%
Fat	100 g	Maize (corn)	5%
Vitamin pre-mix	5 g	Bone meal	10%
D,L-Methionine	1.43 g	Shrimp meal	5%
L-Lysine	4.79 g	Dried milk (spray-dried)	5%
Trace elements	0.75 g	Yeast	2.5%
		Calcium	2%
Analysis:		Salt	1%
		Vitamin pre-mix	0.5%
Crude protein	24.4%	Colouring matter	0.2%
Oil	10%	Binder	0.3%
Fibre	2.5%		
			100%

Feed E:

Analysis:

Fish meal	25%	Crude protein	36.0%
Barley	26.4%	Oil	5.0%
		Fibre	3.5%

Feeding frequency   by hand 10 times a day at hourly intervals

Feed rationing   up to 76th day *ad lib* (demand)
                     from 77th day onwards:

Weight of fish			*Ration*
up to	$\bar{x}$	10 g	10% of liveweight per day
$\bar{x}$ 10 g —	$\bar{x}$	20 g	8% of liveweight per day
$\bar{x}$ 20 g —	$\bar{x}$	30 g	6% of liveweight per day
$\bar{x}$ 30 g —	$\bar{x}$	50 g	5% of liveweight per day
$\bar{x}$ 50 g —	$\bar{x}$	100 g	4% of liveweight per day

Tank   20 l, glass aquaria (76 days);
                     40 l, glass aquaria from 77th day onwards

Water source   spring
Water
temperature   approximately $23°C$

Water flow rate   approximately   0.8 l/min and 1.5 l/min from 77th day onwards
Weighing
frequency   weekly; fish weighed individually every 4 weeks

SUMMARY OF DATA AND RESULTS

Experiment code	73/8—1	73/8—2	73/8—3	73/8—4	73/8—5
Feed	E	SM3	68% SM3	68% SM3	Algae
			32% Lucerne	32% Algae	*Scenedesmus*
		F	G	H	I
At start					
Σ Weight (g)	11.95	12.1	12.05	11.9	11.95
No. of fish	20	20	20	20	20
$\bar{x}$ Weight (g)	0.6	0.6	0.6	0.6	0.6
At end					
Σ Weight (g)	175.9	608.7	509.3	1318.4	392.3
No. of fish	20	19	20	20	20
$\bar{x}$ Weight (g)	8.8	32.04	25.47	65.92	20.65
Σ Weight increase (g)	163.95	597.2	497.25	1306.5	380.95
$\bar{x}$ Weight increase (g)	8.2	31.44	24.87	65.32	20.05
Conversion ratio	15.61	5.75	6.05	2.84	6.75

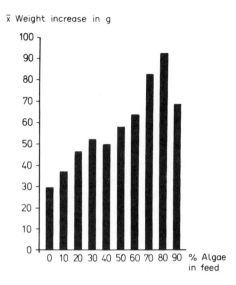

Fig. 96: Weight increases in grass carp on fish meal feed containing from 0% to 90% green algae. Conversion ratios are shown on top of the columns. (Experiment 73/17.)

Fig. 97: Sterlets (*Acipenser ruthenus*) in an aquarium.

ASCERTAINING THE OPTIMAL INCLUSION RATE OF GREEN ALGAE (Scenedesmus) IN GRASS CARP FEED
EXPERIMENT 73/17

In this experiment there was a progressive improvement in growth and feed conversion in the fish as the algal content was increased from 0% to 80% (Meske and Pruss, 1977). This is shown in the table below and illustrated in Fig. 96. The fish fed 80% algae weighed about three times as much as the fish on a diet without algae at the end of the 4-month-long experiment. Raising the algal inclusion to 90% reduced growth, presumably because of deficiencies.

Considering the results of these feed trials it is clear that with 80% algal inclusion a very good conversion ratio of 1.34 is achievable in grass carp. Similar results have been described by Huismann (1978) using feeds containing fish meal. Antalfi and Toelg (1971) report a conversion ratio of between 1 to 20 and 1 to 70 when feeding grass carp on fresh aquatic plants, and Mann (1968) found a ratio ranging between 1 to 30 and 1 to 43 for fresh lettuce.

In recent experiments poor growth was observed when grass carp was suddenly switched to a diet of grass only after having been fed for 1 year on trout concentrate containing fish meal.

The aim of future research should be the investigation into the feeding of grass carp with suitable non-aquatic plants. On the basis of the experiments described here it is unlikely that grass carp can be produced on feed containing no fish meal. This view is shared by Fischer (1973) and Huismann (1978). Nevertheless a substantial component of grass carp feed could be of plant origin. This may mean that suitable plant crops need to be specifically grown for grass carp. Algal production is one possibility, not only under suitable natural climatic conditions but also in a controlled environment. Another is the cultivation of suitable forage crops.

MATERIAL AND METHOD

Duration	119 days
Fish	grass carp from mating 3046
Treatment	73/17— 1 + 2: 20 grass carp each on 100% Feed F
	73/17— 3 + 4: 20 grass carp each on 90% Feed F and 10% algae*
	73/17— 5 + 6: 20 grass carp each on 80% Feed F and 20% algae
	73/17— 7 + 8: 20 grass carp each on 70% Feed F and 30% algae
	73/17— 9 + 10: 20 grass carp each on 60% Feed F and 40% algae
	73/17—11 + 12: 20 grass carp each on 50% Feed F and 50% algae
	73/17—13 + 14: 20 grass carp each on 40% Feed F and 60% algae
	73/17—15 + 16: 20 grass carp each on 30% Feed F and 70% algae
	73/17—17 + 18: 20 grass carp each on 20% Feed F and 80% algae
	73/17—19 + 20: 20 grass carp each on 10% Feed F and 90% algae
Feed preparation	Feed and mixtures were moistened with water, put through a mincer, the skeins dried and then crumbled to required size

*   Roller-dried algae.

Tank	40 l, glass aquaria
Water flow rate	approximately 2.5 l/min
Water temperature	24°C
Feeding frequency	10 times a day at hourly intervals
Feed rationing	*ad lib* for first 49 days
	from 50th day onwards as Table 5, page 51

RESULTS OF FEED TRIALS

Experiment code No.	Percentage of algae in feed	Average weight of fish at end of experiment	Food conversion ratio
73/17− 1 + 2	0	32.12	3.04
73/17− 3 + 4	10	39.92	2.36
73/17− 5 + 6	20	49.17	2.04
73/17− 7 + 8	30	54.71	1.88
73/17− 9 + 10	40	52.37	1.80
73/17−11 + 12	50	60.48	1.73
73/17−13 + 14	60	66.48	1.57
73/17−15 + 16	70	85.73	1.42
73/17−17 + 18	80	95.60	1.34
73/17−19 + 20	90	71.77	1.69

INDUCED SPAWNING

The artificial propagation of grass carp in warm water is similar to the method of hypophysation described for the carp. Antalfi and Toelg (1971) developed a practical method of artificial ovulation in the plant-eating cyprinidae which they demonstrated at Ahrensburg in 1973, making it possible to raise artificially spawned grass carp on many occasions.

They also established a correlation between the girth of the female and the dosage of acetone dried pituitary extract with which it needs to be injected (Table 14). The average weight of a dried carp pituitary is about 3 mg. The sex of the grass carp is relatively easy to establish because the mature male exhibits rough pectoral fins and the female has a larger and more rounded belly.

TABLE 14

Max. girth of female (cm)	38	42	46	50	54	58	62
Dosage (mg/kg liveweight)	3.0	3.5	4.0	4.5	5.0	5.5	6.0

The treatment is in three stages:

1. The female is tranquillized with an anaesthetic such as Sandoz MS-222 and its maximum girth measured to establish the dosage.

2. Pre-treatment of the female with 10% of the dosage (fractional injection). The dried pituitary is ground and then suspended in a physiological saline solution which is injected into the dorsal musculature. Males do not need pre-treatment, but both the male and female fish should now be put into a larger tank so that they can display.

3. Twenty four hours after the pre-treatment the females are injected with the rest of the dosage and the males are given 9 mg irrespective of their girth or weight.

If the treatment is successful the fish will start their display about 10—12 hours after the female's last injection. The male will attempt to envelop the posterior part of the female with its caudal fin. When activities reach their peak, or at the latest when foam bubbles appear prior to milt ejaculation, the fish are carefully removed from the spawning tank and stripped. The eggs, followed immediately by the milt, are collected in a plastic bowl and dry-mixed at once with a feather, or by rocking and panning of the bowl. Mixing is followed by the repeated adding of clean water without any chemical additives and decanting. The now fertile eggs swell considerably. Finally the eggs are placed in a Zuger jar where they remain until hatching with a continuous water flow being maintained.

Because they swell considerably pelagic eggs require a lot of space. Hence an adequate number of jars should be prepared before hypophysation commences. Larger vessels, which can be plastic, can be used provided they are similar in shape to the Zuger jars and allow constant water circulation.

If the water temperature is kept at 23°C the eggs will hatch in 2—3 days and are best fed with *Artemia salina* larvae. After about 10—14 days fine concentrate meal can be introduced to the feed. Trout fry concentrate is suitable for the next stage. Experiments on the feeding of young grass carp are described above.

STURGEON

Members of the sturgeon family are extremely sensitive to pollution. As industrialization of western Europe progressed they were almost wiped out. From 1892 to 1918 the annual catch of sturgeon in the German bight decreased from 4896 to 34 fish per year (Knoesche, 1969). Even in the Soviet Union, which accounts at present for 90% of the global sturgeon population (Schlotfeldt, 1971), catches have been greatly reduced because of pollution. The sturgeon population in the USSR has increased during the last few years because of the development of pond husbandry techniques.

Hybrids of sterlet (*Acipenser ruthenus*) x beluga (*Huso huso*) are of interest and some economic importance. They are obtained by artificial fertilization and they are fertile because the two species both have 60 chromosomes (Nikoljukin, 1966).

These so-called pond sturgeon distinguish themselves by their good growth rate and early sexual maturity (Merla, 1970). The eggs are obtained by Caesarian section, the abdomen being stitched up after the operation (Burzev, 1969).

The Ahrensburg Institute was able to obtain some young sterlets (Fig. 97) after prolonged and persistent efforts. They were kept at a water temperature of 23°C and fed predominantly on gnat larvae. These were taken up immediately whereas fish meat and a concentrate were often left untouched. The growth of the fish was variable. After some losses the remaining 12 sterlets grew on average from 33.8 g each to a maximum of 135 g in just under 3 months.

THE EUROPEAN WELS OR SHEATFISH (*Silurus glanis*)

MANAGEMENT

The intensive rearing of the European wels or sheatfish (also described as a silurid catfish) poses some problems. The fish tends to be aggressive when feed is restricted. Biting then occurs, resulting in injuries, particularly to the pectoral fins. Population density is obviously the decisive factor which triggers off this behaviour. Observations have shown, however, that with very high densities aggression is inhibited. It has also been observed that young fish may turn to cannibalism during fights for dominance.

In its natural environment the wels is a bottom-dweller which shuns light. During the day it stalks its prey using aquatic weeds or river and lake banks as cover. The experiments have shown that the wels can be successfully cultivated either in blacked-out tanks or in glass aquaria with equally good results. Wels fed on dry feed adapt to daylight activity provided they are fed during the day. If fed on live carp fry the fish will only feed during the night even when kept in glass aquaria. If the wels is to be commercially farmed it is necessary to obtain data on its management. The possible commercial exploitation of this species is the reason that research on the wels is now being undertaken at Ahrensburg. Interim results indicate that, against expectations, the wels achieve good growth in conical containers whose sides are blacked out, with light only able to enter from the surface.

FEEDING

The carnivorous *Silurus glanis* can be fed on protein of various origins. Depending on the size of the fish being studied they were fed carp fry and fingerlings, minced frozen fish meat, offal and dry trout concentrate. However it has not yet been possible to carry out large-scale experiments with specific feeds which would give clear and significant results. The aim of the experiments listed here was primarily to ascertain whether the European wels can make a contribution to commercial table fish production when it is cultured in warm water at 23°C.

Table 15 shows the results of experiment 70/87. It gives the individual weights of nine wels over a period of 2 years. At the start of the experiment their weight averaged 25.8 g. At the end their weight varied widely, averaging 2416.3 g with the heaviest fish weighing nearly 4 kg. The growth potential of the European wels is, however, considered much greater.

Following artificial ovulation at Ahrensburg a wels hatched on 26 May 1973 and kept at 23°C grew to an extent unparalleled in European fish and only matched by the pirarucu (*Arapaima gigas*), one of the world's largest fresh water fish (Bardach *et al.*, 1972) which is discussed below. The performance of this wels is summarized in Table 16 (Experiment 74/27) giving its weight at roughly ½-yearly intervals. At 3 years the fish weighed over 17 kg and after 5 years

TABLE 15

Data of Experiment 70/87 (individual weights, in grams)

S. glanis growth rates

Date	Fish code Nos.								
	3	4	5	6	8	9	10	11	14
07.12.70	22	31	26	24	31	39	18	20	21
14.01.71	55	65	75	75	110	145	80	40	50
12.02.71	61	95	91	85	150	185	105	64	62
16.03.71	75	120	110	110	160	190	120	50	60
15.04.71	105	165	155	150	170	270	150	75	80
07.05.71	115	160	185	155	170	280	165	70	75
07.06.71	155	210	230	210	200	395	230	125	115
03.07.71	175	265	290	290	210	410	280	120	150
26.08.71	220	390	455	335	270	670	370	170	175
29.09.71	225	440	570	335	310	900	525	170	200
21.10.71	365	450	790	360	360	1135	665	170	240
19.11.71	545	505	890	550	530	1270	820	210	360
15.12.71	730	690	1150	750	705	1370	1060	315	500
19.01.72	875	1030	1335	770	785	1850	1200	470	575
15.02.72	910	1180	1490	795	875	2435	1175	465	605
09.03.72	935	1090	1440	770	850	2535	1080	440	605
05.04.72	990	1025	1165	830	890	2415	1395	460	600
09.06.72	1350	1360	1800	1240	1410	2730	1570	600	710
03.07.72	1535	1570	1960	1230	1250	2890	1865	545	725
15.12.72	1935	2870	3130	1625	1390	3920	3415	1045	—

TABLE 16

Data of Experiment 74/27: Weight of one individual European wels
(*Silurus glanis*) ex cross 3045, hatched 26 May 1973

Date	Weight (g)
26.05.73	0
20.06.74	2550
16.12.74	5000
23.06.75	7785
09.12.75	12 400
16.06.76	17 300
20.12.76	23 300
12.07.77	24 800
12.12.77	29 100
28.06.78	31 800
17.10.79	40 000

nearly 32 kg. Up to weight of 5150 g this fish was fed dry concentrate only. After that it was given live young carp.

Figure 98 shows the fish in its 2 m long aquarium when it weighed 24.8 kg and had a length of 1.25 m. The fish became the centre of attraction at agricultural shows, and although the press called it Max it was a female.

The remarkable growth performance of this wels demonstrates the possible potential of *Silurus glanis* for warm water farming. Sheatfish caught at dams in Czechoslovakia averaged 1.9 kg at 5 years and reached 33 kg at 20 years (Hochmann, 1966).

Other Ahrensburg experiments involving several group of wels in aquaria at 23°C were carried out and good growth observed. One group of eight fish fed trout meal exclusively started with an average weight of 65 g in December 1975 and attained an average weight of 6.2 kg in July 1980.

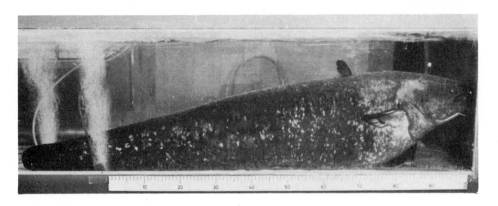

Fig. 98: European wels (*Silurus glanis*) raised in an aquarium, when 4 years old.

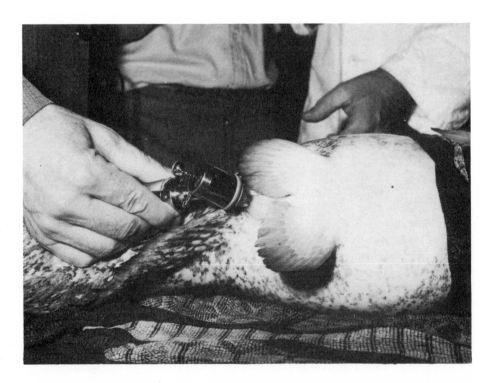

Fig. 99: Sexing the European wels or sheatfish (*Silurus glanis*) with the aid of an auroscope.

BREEDING

Artificial ovulation of the wels through hypophysation is possible and is practised in Hungary and Yugoslavia in particular. The problem is sexing: male and female wels look alike. There is a slight difference in the genital openings but this is not always sufficient for a positive identification. Fijan (1973) uses an auroscope which allows inspection of the internal sexual organs of an anaesthetized wels. Apart from establishing the sex of the fish the stage of maturity of the gonads can also be ascertained by this method.

Figures 99–101 illustrate stages of artificial insemination in the European wels as demonstrated by Fijan in 1973 at Ahrensburg. This procedure has been successfully repeated on numerous occasions.

Initially females are injected intramuscularly with 3 mg/kg liveweight of dried carp pituitary in physiological saline. After 24 hours they are given another dose of 4 mg/kg liveweight. After about 14 hours stripping can commence. This is not always easy with large fish (Fig. 100). Obtaining milt from mature males is even more difficult. On injection of 3 mg of dried carp pituitary extract per kg of liveweight the fish often do not ejaculate on pressure being applied. In such cases the fish are killed and the gonads cut into fine pieces with scissors to be mixed with the eggs. After a few minutes Woyarovich's fertilizing fluid of 40 g NaCl and 10 g carbamide in 10 l of water is slowly added. It is useful to deposit the eggs on a sloping gauze frame (Fig. 101) placed in a dish which is irrigated by a slow flow of warm water (Fijan, 1973).

It is also possible to hatch the eggs in large Zuger jars (Horvath, 1977). To prevent the eggs sticking together an enzyme is added for a short period (Horvath, 1979). Flushing briefly with 5 mg/l of malachite green fungicide is recommended. Kept at 23°C the eggs will hatch in about 3 days (Fig. 102).

Initially the newly hatched wels is best fed brine shrimp (*Artemia salina*) larvae. Dry concentrates, such as a salmon starter, can be used and have been successfully employed at Ahrensburg.

In addition to the work on the European wels experiments with the channel catfish (*Ictalurus puntatus*) have been carried out since 1969 to establish the contribution this fish could make to aquaculture when kept at 23°C. Compared with the European wels kept at that temperature the growth of the channel catfish was found to be slower. Hilge (1978), working on the channel catfish at Ahrensburg, describes its growth as rather slow at 23–25°C, the fish under these conditions taking 3 months from 10 g to 30 g liveweight. In 1980 he reported further experiments carried out under the same conditions in which growth from 10 g to 1 kg took between 15 and 21 months.

TILAPIA spp

In addition to the European fish which respond to warm water culture, non-native table fish should also be considered for production providing their management, feeding and breeding can be suitably controlled.

Several species of tilapia meet these conditions and a start was therefore made at Ahrensburg to test these fish:

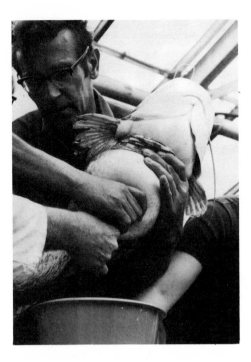

Fig. 100: Stripping a European wels at the Ahrensburg warm water installation.

Fig. 101: Fertilized wels eggs being placed on a gauze frame.

Fig. 102: Fry of the European wels (*Silurus glanis*).

Fig. 103: Male *Tilapia nilotica* bred in warm water closed-circuit installation at Ahrensburg.

the Nile tilapia,   *Tilapia nilotica* (Fig. 103)

the golden tilapia,   *Tilapia aurea*

the Java tilapia,   *Tilapia mossambica*

the Galilee cichlid   *Tilapia galilaea*

These species are all mouth breeders and have, as explained on page 15, been re-classified as *Oreochromis niloticus, O. aureus, O. mossambicus* and *Sarotherodon galilaeus* respectively.

MANAGEMENT

It is possible to keep the above tilapia species within the confines of aquaria with fresh water through-flow and at a temperature of 23°C or more if possible.

Their aggressive behaviour, however, presents problems not found in carp and grass carp culture. Mature tilapia can establish such dominance in an aquarium population that the other fish hardly feed because of continuous attacks. The attacked fish die. This was particularly and repeatedly observed in 10 fish groups of *T. mariae* kept in 40 l aquaria. After a few weeks only one male survived in each group, having attacked all the other fish persistently until they had died.

The aggressive behaviour as observed at Ahrensburg occurs when these fish are kept in small containers (20—80 l aquaria) and when they are kept at low stocking density. The aggression can be contained by using large tanks with a capacity of 500 l or more. Alternatively higher stock density even in small-volume tanks can be used to combat aggression.

Experiment 79/1 describes the growth rate and losses in 40 l aquaria (Fig. 104).

THE INFLUENCE OF STOCKING DENSITY ON TILAPIA
EXPERIMENT 79/1

The results of the experiment show clearly the difficulties which can arise in work with tilapia. There is a danger that the answers to certain questions — such as the efficacy of specific feeds — may be influenced by behaviour problems peculiar to tilapia.

In this experiment only 20% of the fish in the low-density tanks (five fish per 40 l aquarium) survived. In the case of the high stock density tanks (20 tilapia per 40 l aquarium), however, nearly 99% of the fish survived. Further, the average weight of the fish in the densely stocked aquaria was at 103 g, over double that in the low-density aquaria (48 g).

The aquaria all had the same volume (40 l) but the water flow rate was raised commensurate with the stocking rate (from 0.5 to 3 l/min). Since survival and average weights were considerably superior at the higher than at the lower stocking rate, the explanation must be sought in terms of the behaviour of this species. At lower stocking density a higher degree of aggression was observed. As a result the fish lower down the hierarchy suffered severe stress, stopped feeding and often died. At the high density — with more than 3 kg of fish in the 40 l aquaria at the end of the experiment – aggression did not appear to develop and in spite or because of the crowding performed considerably better.

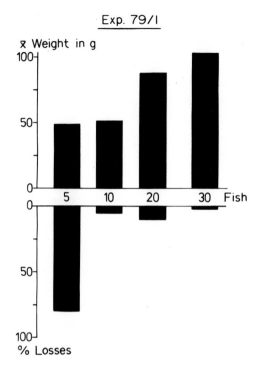

Exp. 79/1

Fig. 104: The influence of stocking density on tilapia (Experiment 79/1). Growth rate and losses in 40 l aquaria.

Fig. 105: The influence of temperature on growth in *Tilapia galilaea* (Experiment 82/7). Figures above columns are the feed conversion ratios.

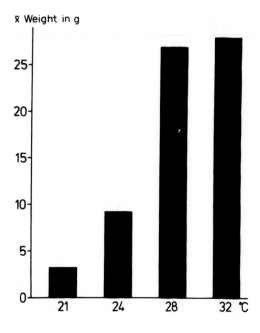

Exp. 82/7

MATERIAL AND METHOD

Duration	97 days		
Fish	*Tilapia aurea* (mating 8036/37)		

Programme		No. of fish	Water flow rate (l/min)	
	79/1−1:	5	0.5	)
	79/1−2:	10	1	) 1st replication
	79/1−3:	20	2	)
	79/1−4:	30	3	)
	79/1−5:	5	0.5	)
	79/1−6:	10	1	) 2nd replication
	79/1−7:	20	2	)
	79/1−8:	30	3	)

Feeding regime	− 10 g	10% of liveweight per day
	10 − 20 g	8%
	20 − 30 g	6%
	30 − 50 g	5%
	50 − 100 g	4%
	100 − 600 g	3.5%

Feed	trout concentrate. Analysis: protein 47%, oil 8%, fibre 3.5%, ash 10%
Tank	40 l, glass aquaria
Water temperature	25°C
Water source	direct through-flow, from well
Water flow rate	see programme above

SUMMARY OF DATA AND RESULTS

Code No. of fish per aquarium	79/1−1 + 5 5	79/1−2 + 6 10	79/1−3 + 7 20	79/1−4 + 8 30
At start				
No. of fish	10	20	40	60
Σ Weight (g)	97.48	188.32	373.72	578.88
$\bar{x}$ Weight (g)	9.75	9.42	9.34	9.65
At end				
No. of fish	2	19	36	59
Σ Weight (g)	96.80	987.98	3181.96	6097.69
$\bar{x}$ Weight (g)	48.40	52.00	88.39	103.35
Weight increase (g)	− 0.68	799.66	2808.24	5518.81
Weight increase (%)	0.7	424.63	751.43	953.36
FCR	−	4.32	2.74	2.45

THE INFLUENCE OF TEMPERATURE ON GROWTH IN TILAPIA
EXPERIMENT 82/7

If tilapia species are to be exploited commercially in non-tropical regions, using for example waste heat from power stations, it is essential to ascertain the most suitable water temperature for growing these fish.

In experiment 82/7 *Tilapia galilaea* were kept at 20, 24, 28 and 32°C respectively. The results, illustrated in Fig. 105, indicate clearly that water temperature must be kept above $24°C$ for optimal results, i.e. above the temperature used in the production of other warm water table fish such as the carp, the eel and the wels.

The best feed conversion at 1.61 was achieved in this experiment at 28°C. At 32°C conversion was less efficient, the conversion ratio being 1.74.

MATERIAL AND METHOD

Duration	105 days
Fish	*Tilapia galilaea* of X 2002
Programme	82/7—1 + 2 kept at approx. $20°C$
	82/7—3 + 4 kept at approx. $24°C$
	82/7—5 + 6 kept at approx. $28°C$
	82/7—7 + 8 kept at approx. $32°C$
	30 fish per tank
Tank	20 l, glass aquaria
Water flow rate	approximately 1 l/min
Water source	well
Feed	fine trout concentrate meal. Analysis: protein 47%, oil 8%, fibre 3.5%, ash 10%
Feeding rate	5% of liveweight per day
Feeding frequency	5 times daily
Weighing	weekly

DATA AND RESULTS

Experiment code no.	Water temperature (approx.) (°C)	No. and weight of fish						Weight increase		Feed fed	FCR
		At start			At end						
		No.	Total weight (g)	Mean weight (g)	No.	Total weight (g)	Mean weight (g)	Total (g)	(%)		
1 + 2	21	60	52.85	0.88	44	141.83	3.22	144.90	274.2	597.8	4.13
3 + 4	24	60	52.79	0.88	60	557.62	9.29	504.83	956.3	1281.5	2.54
5 + 6	28	60	53.14	0.89	60	1624.21	27.07	1571.07	2956.5	2528.4	1.61
7 + 8	32	60	53.05	0.88	60	1690.13	28.17	637.08	3085.9	2846.8	1.74

TANK SHAPE AND BREEDING BEHAVIOUR IN TILAPIA
EXPERIMENT 82/1

Uncontrolled breeding is a problem with intensively managed tilapia (*Oreochromis*) species. Most members of these species display specific generic breeding behaviour. *T. niloticus* males, for example, will continue to attempt to prepare a nest even on the bottom of a smooth glass aquarium and in absence of any substrate. They jealously guard their territory and at best tolerate a female. The courtship and breeding patterns of these fish have considerable influence on efficient food production.

In experiment 82/1 various tank shapes were investigated to see whether shape could be a useful managerial aid to tilapia production. Four types of tank, as shown in Fig. 106, were tested: circular, oval, rectangular, and a cylinder with a conical bottom. It was thought that the latter may be of some interest in the management of these fish. Each tank had a volume of 700 l.

The highest weight increases were achieved in the cone-bottomed cylinder, as can be seen from the tabulated results and Fig. 106. This is presumably due to the breeding behaviour being frustrated by the sloping sides of the cone. In the flat-bottomed tanks eggs and fry were found and there was less growth compared with the coned cylinder.

It would appear that it may be possible to override the behaviour of these fish through managerial measures.

MATERIAL AND METHOD

Duration   258 days

Fish   *T. nilotica* crosses

Programme   82/1—1 + 2   cylinder with cone-shaped bottom
                    82/1—3 + 4   oval tank
                    82/1—5 + 6   rectangular tank
                    82/1—7 + 8   circular tank

                    each tank contained 70 fish

Tank volume   700 l approximately

Feed   carp concentrate. Analysis: protein 50%, oil 6%, Fibre 5.5%

Feeding scale   1% of liveweight per day

Feeding method   manual, 5 times daily

Water flow rate   27 l/min approximately

Water temperature   first 92 days: 24°C; from 93rd day onwards: 30°C

DATA AND RESULTS

Experiment code no.	Tank shape (all 700 l) (Fig. 106)	No. and weight of fish						Weight Increase		Feed fed (g)	FCR
		At start			At end						
		No.	Total weight (g)	Mean weight (g)	No.	Total weight (g)	Mean weight (g)	(g)	(%)		
1 + 2	cylinder, cone-bottomed	140	51 097	365.0	136	99 563	732.1	50 588	99.00	206 163	4.08
3 + 4	oval	140	51 107	365.1	136	90 812	667.7	41 655	81.51	196 332	4.71
5 + 6	oblong rectangular	140	51 099	365.0	137	82 464	601.9	32 957	64.50	186 018	5.64
7 + 8	circular	140	51 100	365.0	132	72 468	549.0	25 081	49.08	176 787	7.05

ADAPTATION TO INCREASING SALINITY IN TILAPIA
EXPERIMENT 78/15: **Nile tilapia (Tilapia nilotica)**
EXPERIMENT 78/14: **Tilapia hybrids (T. nilotica x T. aurea)**

During the development of a closed salt-water circuit system tilapia were kept at various levels of salinity. According to reports from several authors it is possible to keep Java tilapia (*Tilapia mossambica*) in both brackish and sea water (see Kirk, 1972). Brock (1954) states that this species will spawn in nearly 35‰ salinity. On the other hand Uchida and King (1962) have shown that mortality of Java tilapia rises with salinity. Chervinski (1961) found no difference in the growth rate for Nile tilapia in fresh or salt water.

The work at Ahrensburg gave similar results. It is most important to acclimatize the fish slowly to salinity. It was possible to take Java and Nile tilapia as well as their crosses from fresh water to 28‰ salinity in stages over 7 days without losses. The following experiments describe the effect of the rate at which salinity is raised on the mortality of some tilapia species.

ADAPTATION TO INCREASING SALINITY BY NILE TILAPIA
EXPERIMENT 78/15

The Nile tilapia in this experiment were unable to adapt to the gradual increase in salinity in 3 and 5 days respectively. No losses occurred, however, when the salinity was raised from 0% to 28‰ over a period of 7 days.

MATERIAL AND METHOD

Duration	10 days
Fish	Nile tilapia mating 8025—28
Method	3 aquaria containing 10 fish each of 37.6 g $\bar{x}$ weight

	Salt added in g per 60 l per day Aquarium No.			Salt content in g/l (resulting salinity) Aquarium No.		
Day	1	2	3	1	2	3
1	560	336	240	9.33	5.6	4.0
2	560	336	240	18.66	11.2	8.0
3	560	336	240	28.00	16.8	12.0
4	—	336	240	28.00	22.4	16.0
5	—	336	240	28.00	28.0	20.0
6	—	—	240	28.00	28.0	24.0
7	—	—	240	28.00	28.0	28.0
8—10	—	—	—	28.00	28.0	28.0

Water	stagnant, temperature 21—22°C. Water changed every second day at previous salinity and temperature

DATA AND RESULTS

Fatalities each day of the experiment and eventual survivors.

	Aquarium no.		
Day	1	2	3
1	0	0	0
2	0	0	0
3	8	0	0
4	1	6	0
5—10	0	0	0
Survivors at end of experiment	1	4	10

ADAPTATION TO RISING SALINITY IN TILAPIA HYBRIDS (T. nilotica x T. aurea)
EXPERIMENT 78/14

Switching these tilapia hybrids (*T. nilotica* x *T. aurea*) directly from fresh to salt water of 28 g/l proved fatal to these fish fairly quickly. Uchida and King (1962) reported that even after successful adaptation the mortality rate persisted over a period of months at salinity levels over 30%oo. However, a satisfactory growth rate was maintained.

The experiments appear not to promise much success for commercial tilapia production in water with high levels of salinity.

MATERIAL AND METHOD

Duration   10 days

Fish   Tilapia hybrids (*T. nilotica* x *T. aurea*)

Method   5 x 60 l aquaria each containing 10 fish of 14.5 g $\bar{x}$ weight

	Salt added in g to each 60 l aquarium per day Aquarium No.				Salt content in g/l (resulting salinity) Aquarium No.			
Day	1	2	3	4	1	2	3	4
1	1680	560	336	240	28.0	9.33	5.6	4.0
2	—	560	336	240	28.0	18.66	11.2	8.0
3	—	560	336	240	28.0	18.00	16.8	12.0
4	—	—	336	240	28.0	18.00	22.4	16.0
5	—	—	336	240	28.0	18.00	28.0	20.0
6	—	—	—	240	28.0	18.00	28.0	24.0
7	—	—	—	240	28.0	18.00	28.0	28.0
8–10	—	—	—	—	28.0	18.00	28.0	28.0

Water   stagnant, at 21–22°C. Changed every second day at previous salinity and temperature

DATA AND RESULTS

Number of fatalities on each day of experiment, and number of survivors.

	Aquarium			
Day	1	2	3	4
1	0	0	0	0
2	7	0	0	0
3	2	0	0	0
4	1	0	0	0
5–10	0	0	0	0
Survivors at end of experiment	0	10	10	10

FEEDING

Several tilapia species are predominantly omnivorous (e.g. *T. mossambica*) while others are herbivores by preference (*T. nilotica*). The feed range of tilapia generally is fairly wide. In warm climates there is no problem with farmed tilapia being fed a mixture of plant residues and detritus while also being carnivorous. When kept intensively in tanks they have no access to natural food and a balanced concentrate must be provided. Some work on the Java tilapia (*T. mossambica*) has already been done. Mironova (1974) and Rajami and Job (1976) have reported on its energy balance. Antsyshkina *et al.* (1968) and Mathavan (1974) wrote on the utilization of food from plant and animal sources. Nevertheless a lot of work still needs to be done in this field, with particular emphasis being placed on the basic nutritional requirements of the fish.

Experimenting in aquaria, Cruz and Laudencia (1978) achieved the best growth in Nile tilapia when feeding a 36% protein formulation containing fish meal. Jauncey (1982) obtained the best growth in young *T. mossambica* using a 40% protein feed containing a high percentage of fish meal.

In the Ahrensburg experiments the tilapia were mostly fed on commercial granulated or pelleted trout concentrates. These feeds however contain no green plant matter, such as algae, which in a natural environment would inevitably be taken up.

The question of feeding tilapia in intensive aquaculture systems has not yet been satisfactorily answered. The results achieved with trout and carp concentrates lag far behind those recorded in natural waters. Therefore work continues at Ahrensburg on the subject of feeding intensively managed tilapia.

FEEDING NEWLY-HATCHED *Tilapia galilaea* FRY
EXPERIMENT 82/10

The difficulties met when feeding concentrate to newly hatched cyprinids at Ahrensburg, as at other research centres, have already been discussed. In a similar test the feeding of concentrate, in the form of a fine meal, to newly hatched fry of the chiclid *Tilapia* (*Sarotherodon*) was investigated in a comparison with live brine shrimp (*Artemia salina*) larvae.

Eight 20 l aquaria each containing 100 fry were divided into four groups, A to D. Group A was fed concentrate meal on hatching and throughout the experiment. Groups B, C and D were given brine shrimp for the first 5, 10 and 15 days after hatching respectively. This was followed by 7 days on both brine shrimp and concentrate. After this they were fed concentrate only.

Group A had an 80% mortality rate and little weight gain. There were practically no losses in the groups fed initially on brine shrimp, their growth being related to the length of time they were fed brine shrimp before being switched over to the concentrate (Fig. 107).

In another experiment, two groups, twice replicated, of 200 *T. galilaea* were fed finely ground concentrate and tubifex respectively for 2 weeks after hatching. In this experiment, too, the concentrate resulted in a high mortality rate and little growth, whilst the live-feed group suffered hardly any losses and achieved considerably better growth.

MATERIAL AND METHOD

Duration	49 days
Fish	*Tilapia galilaea* X 2009
Programme	82/10—1 + 2 finely ground concentrate for 22 days

	82/10—3 + 4	*Artemia* for first 5 days, *Artemia* and concentrate for the next 7 days, then concentrate only for 10 days
	82/10—5 + 6	*Artemia* for first 10 days, *Artemia* and concentrate for the next 7 days, then concentrate only for 5 days
	82/10—7 + 8	*Artemia* for the first 15 days, then *Artemia* and concentrate for 7 days

After 22 days all groups were fed concentrate (finely ground) with analysis: protein 47%, oil 8%, fibre 3.5%, ash 10%

each tank contained 100 fish

Water temperature	27.4°C
Water flow rate	1 l/min
Weighing	weekly

DATA AND RESULTS

Experiment code no.	Feeding	No. and weight of fish						Weight increase		Feed fed (approx.) (g)	FCR	Mortality (%)
		At start			At end							
		No.	Total weight (g)	Mean weight (g)	No.	Total weight (g)	Mean weight (g)	(g)	(%)			
1 + 2	finely ground concentrate	200	0.2	0.002	40	53.8	1.35	60.7	30 365	169	2.78	80
3 + 4	5 days on *Artemia*, 7 days on *Artemia* and concentrate, then concentrate	200	0.2	0.002	198	588.5	2.97	588.1	294 050	627	1.07	1
5 + 6	10 days on *Artemia* 7 days on *Artemia* and concentrate, then concentrate	200	0.2	0.002	195	605.4	3.10	605.4	302 695	666	1.10	2.5
7 + 8	15 days on *Artemia* 7 days on *Artemia* and concentrate, then concentrate	200	0.2	0.002	199	684.8	3.44	684.4	342 200	740	1.08	0.5

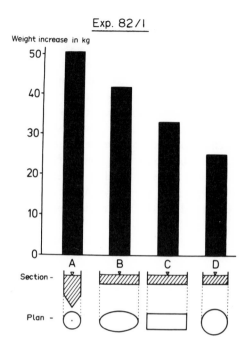

Fig. 106: Tank shape and breeding behaviour in *Tilapia niloticus* (Experiment 82/1). Growth achieved in various types of tanks.

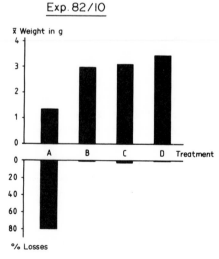

Fig. 107: Feeding *Tilapia galilaea* fry (Experiment 82/10). Newly hatched fish received the following treatments: A: finely ground concentrate meal throughout; B: first 5 days on *Artemia*; C: first 10 days on *Artemia*; D: first 15 days on *Artemia*; followed by 7 days on *Artemia* and concentrate, and then on concentrate only.

BREEDING

Tilapia species are predominantly mouth breeders (e.g. *T. mombassica*). Only a few species are substrate breeders (e.g. *T. zillii*). The mouth breeders hatch their eggs in the mouth of the female where they remain until the yolk sac is absorbed. On hatching the fry will swim closely to the female for a time so that they can return to her mouth when danger threatens.

The procedure at Ahrensburg for breeding tilapia under intensive management is to remove a pair from a collection tank and to place them in a 40 or 80 l glass aquarium. Mature fish ready for spawning are easily detectable because they keep themselves separate from the other fish in the tank. The male instinctively tries to prepare a nest in the substrate but this is not, of course, possible on glass. The male also displays threatening behaviour to rivals invading his territory.

Once placed in the spawning tank the fish tend to spawn within a few days. After fertilization the female commences mouth breeding. It is recommended that the male is removed at this stage.

Only a few hundred eggs can be expected from a female at each spawning. Feeding the free-swimming fry is not difficult. Based on previous experience they were fed brine shrimp nauplii (*Artemia salina*) with satisfactory results. After about 10 days finely ground concentrate can be introduced.

During the work on tilapia attempts were made to strip parent fish of both sperm and eggs, to fertilize the eggs *in vitro* and to hatch them in small Zuger jars. This proved practical but cumbersome. The method has the advantage of shortening the interval between breeding cycles as there is no period of mouth breeding and the female is ready to breed again earlier.

A major handicap of commercial tilapia farming is the ease with which this fish breeds. In Africa, for instance, programmes for pond enterprises are continually threatened by the tilapia's uncontrolled breeding. This results in ponds being overpopulated with small fish. The same problem arises in tanks when there is insufficient control and separation.

One possible solution to the problem lies in mono-sex populations. In tilapia males give better growth than females, but sexing is difficult, time-consuming and inaccurate. It is therefore not a practical option.

However mono-sex populations can be obtained by hybridization of certain tilapia species. As first reported by Hickling (1960) and later by several other authors, including Fishelson, 1962; Pruginin *et al.*, 1975, the progeny of various tilapia hybrids consists almost entirely of males. Some hybrids, e.g. female *T. nilotica* x male *T. aurea*, produced male progeny only. In the majority of the reported hybrids the proportion of males in the progeny lies between 75% and 98%.

Experiments at Ahrensburg have so far produced no improvement in these results. On crossing female *T. nilotica* with male *T. aurea*, males only formed 62% of the progeny while in the female *T. aurea* x male *T. nilotica* cross it was 73% (Klinger, 1977). The evidence now appears to indicate that the production of 100% male populations does not only depend on the species used in the cross but also on their origin, and further fundamental research into this subject is required. Whether this effect is due to genetics or geography needs to be established.

There is another way of producing all male progeny: treating the tilapia fry with hormones. This method uses very small amounts of methyl testosterone administered orally in the feed over a 6-week period, and causes masculinitation in the female fry, resulting effectively in all-male progeny.

The method used at Ahrensburg followed that of Guerrero (1975) on the fry of Nile tilapia (*T. nilotica*) (Klinger and Meske, 1978). 30 mg of methyl testosterone were dissolved in 50 ml of ethyl alcohol and then diluted with distilled water to 1 l. This was admixed to 1 kg of concentrate which was then re-dried.

When the fry reached a length of 9–11 mm (10–20 mg) they were fed the augmented concentrate at the rate of 12% of bodyweight per day for the first 2 weeks of the treatment. During the following 2 weeks they were given the mixture at a rate of 10% of bodyweight and in the final 2 weeks they received it at 8%. After 6 weeks the hormone supplement feeding was replaced by the normal untreated concentrate.

The fish in this treatment were raised to an average weight of 110 g. Their sex was then determined by dissection. The results were as shown in Table 17. The result is a statistically significant predominance of males ($p < 0.001$) but it does not satisfy the expectation of a 100% male population.

TABLE 17

	Fish receiving hormones (124)	Untreated controls (126)
No. of females	43 (35%)	71 (56%)
No. of males	81 (65%)	55 (44%)

PIRARUCU or ARAPAIMA (*Arapaima gigas*) (Figs. 108 and 109)

Arapaima gigas is one of the world's largest fresh water fish. It is known as the pirarucu in Peru (Bardach *et al.*, 1972) while in Brazil it is known as el paiche. It is much sought after as a table fish in the Amazon Basin. It has few secondary bones and a pleasant flavour. According to Sanchez Romero (1961) the *Arapaima* reaches a length of 2.70 m with a weight of 172 kg. Reports of fish 4–5 m long have not been substantiated.

This fish belongs to the Osteoglossidae and is an obligatory air breather. The young fish has to surface 20–30 times in an hour to inhale atmospheric oxygen. Adults need to surface 6–8 times an hour (Schaller and Dorn, 1973). In Europe the first specimen of the pirarucu was raised in an exhibition tank at the London Zoo (Lueling, 1965).

Within the framework of examining exotic fish for their suitability for table fish production in European warm water installations, Ahrensburg was able to obtain some young Arapaimas from Brazil to see whether it would be possible to rear them.

These fish had an average length of 10 cm and an average weight of 35 g at the start of the experiment. They were kept at 28°C in glass aquaria with constant water flow. At the beginning they were fed small live fish, mainly carp. Later their diet was changed to deep frozen trash fish. The growth rate of these arapaimas was impressive (see Table 18). At best the growth of these fish was comparable to that achieved with the European wels (*Silurus glanis*) (see p. 177).

Fig. 108: *Arapaima gigas* or Pirarucu raised in an aquarium from 15 g starting weight. Scale in cm.

Fig. 109: Young *Arapaima gigas* in an aquarium. Scale in cm.

TABLE 18

Growth rate of *Arapaima gigas* (duration of experiment: 310 days)

		$\bar{x}$ Weight (g)	Max. weight (g)
At start		15	19
After	1 month	93	130
	2 months	325	455
	3 months	825	1165
	7 months (approx)	1245	1820
	10 months (approx)	1720	2560

For further details see R. and G. v. Sengbusch, Meske and Cellarius (1974).

Another advantage of *Arapaima* is that among carnivores it displays almost no aggression under intensive management and there is no cannibalism. However experience at Ahrensburg has shown that there are three points which need to be watched. Firstly this fish is extremely lively and given to leaping. Unless there is a suitable cover they easily jump out of their tank. Secondly, because the fish breathes air there is a possibility that they may drown when, for instance, the water level rises because of a blocked drain or when an unsuitable cover denies the fish air. Thirdly, when the water temperature drops below 14°C — through plant malfunction, for example — these fish die because their metabolism is affected and the fish are incapable of surfacing to breathe.

Arapaima would appear to be a suitable fish for warm water aquaculture provided the problems referred to above are taken into consideration and the following points are satisfactorily resolved by further research:

1. formulation of suitable concentrates;

2. the raising to sexual maturity, artificial spawning and controlled rearing of the fry.

As long as the current difficulties of procurement and cost pertain in getting young stock from South America there are limits to the exploitation of this fast-growing and potentially valuable species.

7. THE APPLICATION OF WARM WATER FISH CULTURE IN RESEARCH AND PRACTICE

RESEARCH

The management, feeding and breeding experiments on the carp and other warm water table fish in tanks described in Chapters 5 and 6 are merely the start of planned research programmes which need to be undertaken in this field. Work carried out in closed-circuit systems has the advantage of providing identical environmental conditions in all tanks and, in contrast to experimental work in ponds, allows exact appraisal and complete control of the various factors involved. There are, for example, many outstanding questions on the effects of the different abiotic influences which can be tested with this system. The influence of light intensity and quality on gonadal development in fish is one example. Environmental conditions − such as water temperature, pH, oxygen content, etc. − have a different effect on the various species under discussion. This also applies to poisonous substances: the serious threat of pollution can only be efficiently countered when the effect of toxicity of the various substances on a specific organism is fully established. It is not simply a question of the reaction of the fish to pollutants. At high levels fish may die but at lower levels it is necessary to establish the effect on certain organs and to determine how much of the toxin is absorbed from the water or through the feed. In a controlled environment (e.g. tanks or aquaria) these questions can be systematically investigated. It is also possible, for instance, to test whether the fast-growing and short-lived fish take up more or less poisonous material than a slower-growing fish in cooler water.

The total area available for experiments at Ahrensburg has now grown to about 1000 m^2. It includes glass aquaria of various capacities, channels, circular tanks, silos and special tanks of various shapes. Almost all are connected to the re-circulatory closed-circuit systems. One of the buildings in which experimental work is conducted can be seen in Fig. 110.

Abiotic factors − such as temperature, oxygen and possibly salinity − are of major importance to the efficiency of the purification process as well as to the metabolism and growth of fish. Wild *et al.* (1971) demonstrated the increase in the nitrification rate as the water temperature rose from 5°C to 30°C. Cairns *et al.* (1975) investigated the influence of temperature on chemical toxins and their effect on aquatic organisms. Knoesche (1974) reported on energy budgets of commercial enterprises. Beamish (1964) and Huisman (1976) stressed the important relationship between temperature and oxygen on the metabolism of fish. Rumohr (1975) published a summary on the influence of temperature and salinity on fish growth and sexual maturity.

In recent years there has been a growing number of reports on the husbandry of table fish in warm water installations. Some authors refer to controllable thermal aquaculture and closed-circuit aquaculture (e.g. Yanzito, 1981; Hambrey, 1981; Howell, 1981; Manci and Yanzito, 1982). Results from the use of power station effluent waste heat have been published by Kirk (1972) on tilapia; Sylvester (1975), Aston *et al.* (1976) and Kuhlmann (1979) on the eel; Tanaka (1976) on the yellowtail (*Seriola quinqueradiata*), the Red Sea bream (*Chrysophrys major*) and shrimps, and by Rogers and Cane (1981) on trout. Purdom (1982) wrote on waste heat use.

In these mainly through-flow installations conditions are not always optimal. There may be temperature and salinity fluctuations due to seasonal influences or precipitation. Consistency is more often found in enterprises where the water supply derives from wells. The Rhine-Westphalian Electricity Generation Board plant at Cologne is an example. Here waste heat is used to raise carp, eel, wels and tilapia among others (Stoy, 1981). At the United Kingdom's Central Electricity Generating fish farm at Trawsfynydd, North Wales, 18 t of trout was reared to stocking size in 1 year using waste heat from a nuclear power station circulating in a 500 ha lake at an elevation of 198 m (Rogers and Cane, 1981).

Many experiments with different types of closed-circuit systems involving warm water table fish are being carried out throughout the world (Price, 1978; Mayo, 1979; Neal and Mock, 1979). Sidall (1974) reported on salt and brackish water closed circuits using gravel filters. Forster (1974) reported using upflow filters while Meske and Naegel (1975), Meske and Rakelmann (1980), and Mudrack (1976) describe activated sludge plant systems. V. Westerhagen and Rosenthal (1975) and Rosenthal *et al.* (1979) used an ozonizing method. Ozone is able to destroy pathogenic bacteria of fish (Colberg and Lingg, 1978), eliminate certain viruses (IHNV and IPNV)* (Wedemeyer *et al.*, 1978) and some so-called yellowing substances.

The nitrate content, however, is considerably raised due to N_2 oxides, even in water devoid of nitrogen, by the introduction of ozone with air (Schlesner and Rheinheimer, 1974). The use of ozone in the process of water purification has not proved successful because of the high technical input required, and because of the danger of toxic products being formed.

As regards fresh water circuits, other than the Ahrensburg system, there have been tests with sand filters (Hirayama, 1974), trickling filters (Broussard and Simco, 1976) and also some combined systems (McLarney and Todd, 1977). The Ahrensburg system is described in Chapter 4, in Meske (1979), and Kausch and Ballion-Cusmano (1976), on incorporated denitrification chambers. Tiews (1981) edited many reports on all these subjects.

---

* IHNV; infectious haematopoietic necrosis; IPNV: infectious pancreatic necrosis.

Fig. 110: Part view of one of the buildings in which experiments are carried out in Ahrensburg. Channels can be seen in the foreground with circular tanks further back. In the background are square asbestos cement tanks and glass aquaria. All the tanks are connected to the re-circulatory warm water closed circuit system.

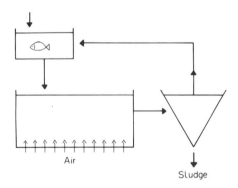

Fig. 111: Water circuit with activated sludge filter (Meske, 1980).

TYPES OF PURIFICATION SYSTEMS

The three main types of purification systems are shown diagrammatically in Figs. 111–113. In these systems purification of the circulating water is achieved entirely by micro-organisms.

ACTIVATED SLUDGE (Fig. 111)

In this system the sludge is aerated to supply the necessary oxygen and to keep the aerobic micro-organisms in continuous movement. The system is described in detail on pages 49–66. Where there is an inadequate supply of oxygen, such as in the sedimentation chamber or in the scum, denitrification takes place by anaerobic bacteria.

The high level of biological activity creates considerable amounts of biomass. In consequence large sludge and sedimentation chambers are required in relation to the fish tank volume served by the plant. The system produces a mainly stabilized surplus of sludge and this, together with the large clearing chambers needed, could be considered a disadvantage.

The great advantage of this type is its long-term proven performance in industrial and domestic sewage treatment. There is also no problem with blockages or clogging of filters. The Ahrensburg system (Chapter 4) is a good example. Its water loss is 2% a day, subject to sludge removal, evaporation and spillage.

UP-FLOW FILTER (Fig. 112)

This type consists of a tank filled with stones or plastic ring elements measuring a few centimetres in diameter and thickness. The water from the fish tanks is first pumped into a pre-cleaning tank, from which the faeces are removed, before flowing up through the filter. The filter media are colonized by bacteria and nitrification and denitrification proceeds in zones which are determined by oxygen availability (see Haug and McCarty, 1972).

The system operates successfully for example in US fish production units where, however, additional denitrification stages are incorporated. There are two drawbacks: the faeces which have to be removed from the system before filtering are an environmental liability and there is no control of the purification process in the filter.

TRICKLING (Fig. 113)

In this system the water trickles over plastic spool-type elements offering large surface areas for the colonization by bacteria. The system requires clearing tanks to remove bacterial matter which is washed off the element from time to time and possibly also for denitrification.

Trickling is an old-established method of purification which is now being superseded by the activated sludge system. Aeration is achieved by the chimney effect without the aid of a compressor. Used in the open these filters need to be insulated, particularly in the case of warm water circuits. Removal of the faeces and sludge may present a pollution problem.

There are also filter bed systems using gravel and sand as biological medium. Where fish are fed concentrates there is the possibility of the filter beds getting clogged up and the faeces then need to be separated out.

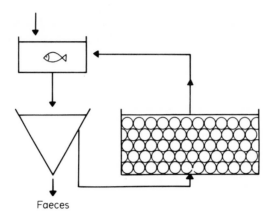

Fig. 112: Water circuit with up-flow filter (Meske, 1980).

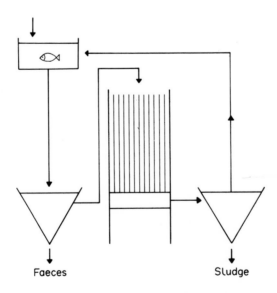

Fig. 113: Water circuit with trickling filter (Meske, 1980).

In general it must be stressed that all filter systems have their limitations and none work satisfactorily all the time. There is loss of water through evaporation, spillage or leakages. In all systems faeces or sludge, or both, have to be removed from the system. Undigested faeces turn into a black, stinking mass which may become a hazard and taking them out of the process only shelves, but does not solve, the problem. The less self-contained the system, the greater the fresh water demand and the threat to the environment.

NUTRITION

Fundamental research into nutrition can only be carried out in aquaria. Firm conclusions cannot be obtained on the utilization of feeds or the fishes' requirements of specific feed components from experiments in ponds. Work in this field has to be carried out to establish the role or effect of every amino acid, certain vitamins and trace elements.

During recent years numerous experiments have been conducted on how best to feed fry. Here the aim is partly to achieve maximum zooplankton production in ponds through the use of fertilizers, and partly on the employment of chemicals which control copepods parasitic on fry (Horvath, 1978). In particular a lot of work was done on the development or formulation of concentrates suitable for cyprinid fry (Seidlitz, 1975; Lieder and Jaehnichen, 1975; Jirasek, 1976; Anwand et al., 1976; von Luckovicz and Rutkowski, 1976). In spite of all efforts there is still no concentrate for cyprinid fry immediately after yolk sac absorbtion. Artemia larvae — or similar — remain the only first feed source (see Chapter 5). As a follow-on to Artemia dry concentrate can be given, and some nutritional values are now available, particularly for carp (Aoe et al., 1970, 1974). Research data on the relation between nutrition and body composition have been published (Kausch and Ballion-Cusmano, 1976; Meske and Pfeffer, 1978b). Halver (1978) reported on vitamin requirements of fish, and Lall (1978), Nose and Arai (1979), Pfeffer and Meske (1978) have reported on the mineral demands.

Much attention is being given to the substitution of fish meal by alternative protein sources for warm water fish (e.g. Viola, 1975; Viola et al., 1981). The search includes the use of dairy by-products (see Chapter 5) and biosynthetic protein (Moebus and Lemke, 1975). Unicellular green algae, such as Chlorella and Scenedesmus, are a natural food source for herbivorous and omnivorous warm water fish, and it is therefore not surprising that their use in feeds often gives good results. A summary of this field was published by Soeder and Binsack (1978). Other authors on the subject are Hepher et al., 1978; Stanley and Jones, 1976; Mironova, 1969. The search for protein extends to experimental feeding of bovine manure to tilapia hybrids. Good growth rates and no tainting was reported (Collis and Smitherman, 1978). Tiews et al. (1981) refers to the considerable body of work concerned with finding substitutes for fish meal in trout feeds.

Spannhof and Kuehle (1977) and Koops and Kuhlmann (1978) have reported the development of a dry concentrate specifically for eels. Nose and Arai (1979) pointed out the essential role of minerals in concentrates for eel growth.

BEHAVIOUR

Behavioural studies on table fish in ponds are more or less impossible. Hence knowledge on this subject is somewhat thin. More research has been undertaken into the behaviour of rare exotic fish of tropical origin than on native European table fish. Courtship display before and during spawning, shoal and hierarchical behaviour are only some of the issues which need clarifying through aquarium experiments. In some species the crowding — i.e. the high stocking density — of intensive husbandry can cause stress which can adversely affect health and growth. Species with distinct courtship display, such as tilapia, are particularly prone to stress, as are voracious carnivores such as the European wels. Peters *et al.* (1980) developed a stress test in which certain blood parameters serve as indicators of stress intensity (see also Sarig and Bejeranao, 1980).

GENETICS AND BREEDING

The very considerable scope for applied genetics and for breeding offered by warm water fish husbandry has already been mentioned in Chapter 5. In these aspects the difference between warm-blooded farm livestock and table fish is particularly noticeable.The opportunities offered by the production and rearing of up to a million progeny from one mating are immense, especially when it is considered that the material can be raised to sexual maturity under completely controllable conditions followed by artificial crossing and hybridization. The capabilities of the present-day laboratory facilitate experiments on gonadal development and reproduction. With the aid of warm water fish husbandry it is possible to standardize environmental conditions to such an extent that their variability can be reduced to insignificance.

A large body of work has recently been concerned with artificial reproduction. Of particular note is the summary by Donaldson (1977) and Chaudhuri (1976) on the use of hormones with fish. Horvath (1978) reports on the relationship between ovulation and water temperature. A variation on the treatment of eggs during controlled multiplication of carp by Woynarovich and Woynarovich (1980) is referred to above. The importance of genetic research for aquaculture is stressed by Moav (1979) and Wilkins (1981). Moav and Wohlfarth (1976) have reported on the work on selection in carp while Bakos (1979) has described hybridization and Kirpichnikov *et al.* (1979) the breeding of disease-resistant fish. In the future three methods of applied genetics promise to be particularly rewarding in the raising of performance in aquaculture: gynogenesis, polyploidy and mutagenesis.

In gynogenesis irradiation is used to inactivate the sperm. The sperm retains the power to penetrate the egg but does not fertilize it. The penetration triggers off embryonic development in the egg. Diploidy is then induced by submitting the female chromosomes to cold temperature shock (Purdom and Lincoln, 1973). In gynogenesis, as in vegetative production, all the genes come from one female source and consequently the progeny is almost entirely female and to a very large extent homozygous (Golovinskaya, 1968; Stanley, 1976; Nagy *et al.*, 1978).

Polyploidy was first achieved by Purdom in flatfish in 1972 when he submitted eggs to cold temperature shock. The result is superior growth performance of these fish. Polyploidy in carp was described by Vasilev *et al.* (1975) and in *Tilapia aurea* by Valenti (1975).

Results on mutagenesis — the induction of mutation by irridiation or chemical treatment were published by Schroeder (1973). A detailed summary on genetics as a basis for effective selection in fish has been published by Kirpichnikov (1981).

To avoid the uncontrolled breeding in tilapia, experiments using two methods to achieve mono-sex populations have been carried out for some years. The aim is to produce only males, because they are the faster-growing sex. One method used involves the crossing of certain tilapia species which results in a high percentage of males (Hickling, 1963; Fishelson, 1962; Pruginin *et al.*, 1975). The other method consists of hormone treatment of tilapia fry. Methyl testosterone is generally used, and should result in an almost complete male progeny (Guerrero, 1975, 1979; Shelton *et al.*, 1978; Anderson and Smitherman, 1978).

Fish kept and fed in warm water throughout the year grow continuously. The growth rate depends on genetic potential and on the many factors which together make up management and feeding (see Chapter 5). An interesting paper by Weatherley (1976) examines the various factors influencing growth in fish. The influence and interrelation of behaviour, population dynamics, endocrinology and feed components is discussed and he stresses that single factors, such as genetics or water temperature, should not be considered in isolation. Overall optimizing of the various factors is difficult. Consequently definitive information on optimal growth is lacking for many fish species. It would appear pertinent to draw attention to the growth rates illustrated in Figs. 114–117 as achieved at Ahrensburg with the carp (*C. carpio*), the grass carp (*C. idella*), the European wels (*S. glanis*) and the European eel (*A. anguilla*). It is possible that these observed growth rates can be improved further. Although the data in the illustrations refer only to the performance of individual fish they can nevertheless serve as a guide to researchers and commercial producers.

SOME PRACTICAL APPLICATIONS

There is a growing appreciation of the possible exploitation of warm water fish culture. In countries with a relatively inclement climate, as for instance in eastern Europe, it is the only form of aquaculture not dependent on ambient temperatures. One way of applying it in practice is to combine it with traditional pond farming. The first experiment in mixed-intensity fish farming commenced at Ahrensburg several years ago (Meske and Schrader, 1968).

A group of summer- and pond-hatched carp were introduced into the warm water installation in September when they weighed 20 g on average. They overwintered at 23°C and continued to feed and grow. In the following May they were placed in two open ponds. Their weight at the time averaged 252.2 g. Also placed in the ponds were carp of approximately the same weight and the same genetic origin hatched a year earlier and therefore 1 year older than the fish which had spent the winter in warm water.

Figure 118 shows the growth of these two groups. The carp overwintered in the warm water continued to grow on release into ponds without check. The carp 1 year younger subsequently overtook the older fish, presumably because of their superior condition at the start of the season after having spent the winter under warm water management.

Overwintering pond carp in warm water and reducing the production cycle by 1 year is only one of the possible management techniques. Spawning, hatching and rearing the fry in a warm water plant can be followed by transfer to open ponds. Experiments were carried out using this technique at Ahrensburg in 1968 and the result is illustrated in Fig. 119. The growth performance observed was outstanding and irrespective of the age of the fish when put into the open ponds. Carp of various age groups were placed in open ponds at the end of May and reached the weights given below during the season:

fish hatched in June of the previous year grew to 2.7 kg;

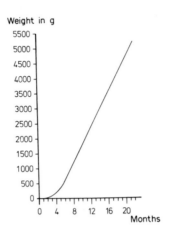

Fig. 114: Carp (*Cyprinus carpio*). Maximum growth observed in aquarium.

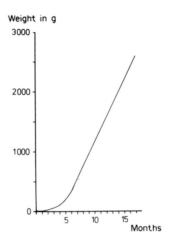

Fig. 115: Grass carp (*Ctenopharyngodon idella*). Maximum growth observed in aquarium.

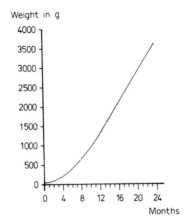

Fig. 116: European wels (*Silurus glanis*). Maximum growth observed in aquarium.

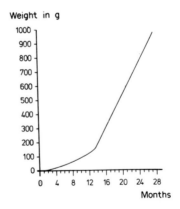

Fig. 117: European eel (*Anguilla anguilla*). Maximum growth observed in aquarium.

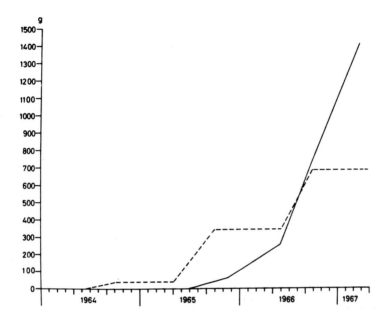

Fig. 118: Average growth rate of carp overwintered in warm water. ———— Growth of fish overwintered in warm water; – – – – – growth of controls in open ponds throughout.

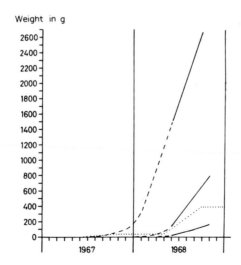

Fig. 119: Growth of carp hatched (on 5.7.67, 7.1.68 and 11.4.68 respectively) and reared in a warm water installation (– – – – – –) when placed in open ponds (————). Dotted line shows growth of pond carp hatched in 1967. Weights are averages.

Fig. 120: Two carp of one pond population photographed 1 year after hatching. The carp at the top was kept in a pond throughout and weighs 40 g. The car at the bottom was placed in warm water installation after hatching and weighs 1750 g.

fish hatched in January of the current year grew to 800 g;

fish hatched in April of the current year grew to 180 g.

Arising from encouraging results a considerable number of carp were produced at Ahrensburg during the early parts of the year for placing in ponds during May on fish farms throughout the various parts of western Germany as a feasibility study (Kossmann, 1970). The growth recorded was predictably variable and ranged between 27 g and 226 g a fish, demonstrating the practical advantages of spawning and rearing of the fish in warm water (Kossman and Szablewski, 1971).

There are now several large fish-farming enterprises in Germany which use their own warm water installations for fry production. In eastern European countries the production and rearing of fry in central warm water plants is now an established practice. As an example the Szazhalombatta installation near Budapest can be cited. Here waste cooling water from a conventional power station is utilized for the mass production of carp, vegetarian cyprinids, European wels and also ornamental exotic aquarium fish. Antalfi and Toelg (1973) justifiably refer to the considerable importance of warm water culture in the planned multiplication and fry rearing.

Taken to its natural conclusion fish would be warm water cultured all the year round. The fish can be managed completely independent of climate or climatic variations and traditional pond farming would have no part in the mode of production. The major advantages of warm water culture are demonstrated in Fig. 120, where a 1-year-old pond carp weighing 40 g is shown alongside a 1.75 kg carp, both being of the same age and originating from the same population. The photograph was taken in June.

Figure 121 demonstrates the uninterrupted growth curve of warm water managed carp which was achieved in an aquarium. Alongside can be seen the growth curve of a pond carp, with its seasonal checks. Both sets of data represent average weights. Warm water culture can accomplish continuous output as required for processing and marketing by hatching batches at monthly intervals. This is illustrated in Fig. 122, from which comparative pond output can also be seen. To get from egg to an 800 g fish in 6 months is possible, as observed at Ahrensburg. Figure 122 demonstrates the potential of warm water installations producing carp monthly through a combination of seasonally independent spawning and fast growth It is 7×800 g = 5.6 kg per year compared with a 30 g output from a pond within the same time span.

Both the through-flow and recirculation or closed-cycle systems can be used in warm water fish farming. In eastern Europe the value of warm water plants has been fairly quickly realized and there is a move to change from traditional management to industrial carp production. Steffens (1979 and 1981) and others have reported on this, production being almost entirely based on through-flow systems using waste heat, particularly from power stations (Babajan, 1974; Vinogradov, 1976; Zobel, 1976).

In (West) Europe there are a number of commercial through-flow installations producing warm water fish for the table. They were developed alongside industrial enterprises, such as dairies, but in the main they are sited at power stations. At the Limnotherm fish farm near Cologne, for example, the current aim is to produce 400 t of fish — eel, tilapia, carp and wels — a year. The farm can draw on 2100 m^3 of cooling water at 19–27°C per hour throughout the year. The present construction programme involves 4200 m^3 of tank capacity. Water is initially

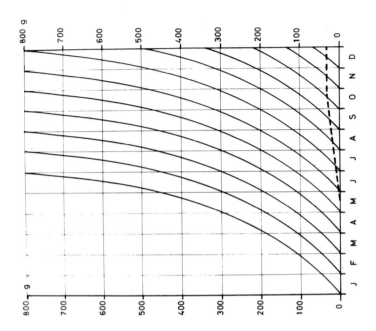

Fig. 122: Projected growth of carp hatched in monthly batches (solid line). Development of pond carp (dotted line) given as comparison.

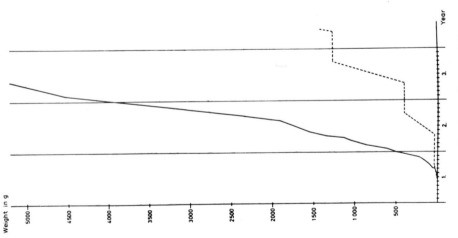

Fig. 121: Growth of carp in central European pond (– – – –) compared with carp in warm water installation (——) during the first 3 years after hatching.

drawn from a well, stabilizing the water parameters (Stoy, 1981). Where surface water is used these parameters tend to be subject to fluctuations. The temperature, for example, may vary with the seasons or during the span of a day. This is discussed by Godfriaux and Stolpe (1981) in their report on the operation of a fish farm at a coal-fired power station in New Jersey, USA. Koops and Kuhlmann (1980) report upon fluctuations in the salinity of cooling water of coastal power stations.

Given suitable conditions it is possible to produce fish native to tropical climates in through-flow installations at power stations. Melard and Philippart (1980) describe the successful production of tilapia (*T. niloticus*) at a nuclear power station on the river Maas in Belgium for 9 months in the year. Aston (1980) discussed the potential for aquaculture at power stations in his survey, referring also to temperature tolerance of the species likely to be of commercial interest and the optimal operating temperatures. Huisman (1981) describes the production of carp (*C. carpio*) and grass carp (*C. idella*) in net cages at a Netherlands power station.

Apart from waste heat, geothermally heated water may also offer possibilities for through-flow plants. An eel farm near Ibusuki in southern Japan uses 40 000 m^3 of geothermally heated, splash aerated, water at 25–30°C a day (Shigueno, 1979). In Idaho, USA, 227 t of channel catfish (*Ictalurus punctatus*) a year are produced in geothermally heated water from an artesian well (Ray, 1981).

Passive and active solar energy may also be exploitable. Zweig *et al.* (1981) and other authors described attempts to set up aquacultural systems producing fish in glass houses. Making the best use of solar energy the primary products here are algae and the secondary crop are golden tilapia (*T. aureus*). MacKay and von Toever (1981) describe an aquacultural re-circulation system under glass on Prince Edward Island, Canada, which includes a hydroponic element. Vogt (1979, 1980, 1981) reports the results achieved with carp in plastic tunnels in the UK. During a 5-month period the fish under plastic grew eight times as fast, ate 64% less food and had a mortality rate reduced from 90% to 30% compared with the open-pond controls. Reference to the work by Yanzito and Manci using active solar energy for their multi-stage radial tank in Wisconsin has already been made (p. 42).

Through-flow installations tend to suffer from the inherent risk of sudden supply failures. During the winter this could lead to the loss of a substantial part of the farm's production. There must also be some apprehension on the partial or total failure in supply due to abrupt incidences of pollution. Re-circulation systems do not suffer from either risk. Power failure would cause the water to cool only gradually without danger to the fish. An abrupt incidence of toxic pollution cannot occur.

The current frequency and intensity of open-water pollution has already been discussed in Chapter 1. It would therefore seem reasonable to incorporate a purifying element in certain waste heat through-flow installations, thereby creating a re-circulation system. In some regions there is now a water shortage so that, in parts of the United Kingdom for instance, fish farms are obliged to re-circulate some of the water in use (Hambrey, 1981). Water shortage is not the only reason for this; there is also the pollution caused by several intensive fish farming enterprises occupying a comparatively short stretch of waterway. In the long term re-circulation is likely to be the most widely used method for aquaculture. This may well also include marine fish.

Closed-cycle systems use more energy, require greater management skills and a higher level of capital investment. Consequently commercial plants producing warm water fish for the table are not common. Further, the systems in practical use are largely based on experience gained through experimental work mainly concerned with cold water fish. Knowledge obtained from cold water systems cannot be readily applied in warm water systems: the purifying micro-organisms display a higher rate of activity at temperatures above 20°C.

A number of studies on the performance of up-flow filters in closed cycle systems are available. Csávás and Varadi (1980) report on Hungarian installations using zeolite gravel. Chiba (1980) describes their application in Japanese eel production while Nemeto (1980) deals with a commercial eel farm, also in Japan. Overall the performances of the up-flow filters in these examples appear not to have been as good as had been anticipated. The flow of water through the filter is not controllable, causing pockets of oxygen deficiency to occur fairly quickly.

Knoesche and Tscheu (1974) have described the advantages of activated sludge systems, and Liao (1980) compares the various closed-cycle systems, suggesting further research and the development of practical, working circuits.

The cost of warm water closed-cycle installations, particularly the high-technology types should be assessed in relation to the alternative or traditional methods of production. In the case of marine fish catches Luther pointed out in 1970 that the cost of a trawler ran to DM 10 million, without there being any guarantee of adequate catches. Vogt (1980) reported that in 1977 the cost of running a vessel of 100 GRT was between £200 and £500 per day (or up to £182 000 p.a.) and for large stern trawlers £2000 to £3000 per day (or £1.1 million p.a.). He also reports that the number of UK trawlers has decreased drastically, and continues to do so, with consequent loss of jobs at sea and ashore. The report also suggests that a fish industry organized along the lines of that of other domestic livestock would provide a not inconsiderable number of permanent jobs and suggests that an annual western European fish crop of 2 Mt of farmed fish is achievable.

A future closed-cycle fish farm may well produce a comparative amount of fish for the same financial input as that expended in sea fishing, except that there is some certainty of a marketable end-product. While sea catches decrease, and there is a deterioration in the quality of pond fish — through industrial pollution, for instance — the unit cost of fish procured by these traditional methods is likely to rise while that of closed-cycle farmed fish must inevitably fall. As others have pointed out, it is perhaps only a question of structuring and organizing the new industry technically and managerially.

REFERENCES

Albrecht, M.L., 1979: Auswirkungen der industriemaeßigen Aufzucht von Karpfen, *Cyprinus carpio*, in Warmwasseranlagen auf die Gonadenentwicklung. *Z. Binnenfischerei DDR*, **26**, H.1, 7–12.

Anas, R.E., 1974: Heavy metals in the Northern Fur Seal, *Callorhinus ursinus* and Harbour Seal, *Phoca vitulina richardi. Fishing Bull.*, **72**, 133.

Anderson, C.E. and Smitherman, R.O., 1978: *Production of Normal Male and Androgen Sex-Reversed* Tilapia aurea *and* T. nilotica *Fed a Commercial Catfish Diet in Ponds.* Symposium on the culture of exotic fishes, Fish Culture Section, American Fisheries Society.

Aoe, H., Ikeda, K. and Saito, T., 1974: Nutrition of Protein in Young Carp — II Nutritive Value of Protein Hydrolyzates. *Bull. of the Jap. Soc. of Scient. Fish.*, **40** (4), 375–379.

Aoe, H., Masuda, I., Abe, T., Saito, T., Toyoda, T. and Kitamura, S., 1970: Nutrition of protein in young carp — I. Nutritive value of free amino acids. *Bull. of the Jap. Soc. of Scient. Fish.*, **40**, H.4.

Aston, R.J., 1980: *The Availability and Quality of Power Station Cooling Water for Aquaculture.* FAO EIFAC/80/Symp., R 3: 1–14.

Aston, R.J., Brown, D.J.A. and Milner, A.G.P., 1976: *Aquaculture at Inland Power Stations.* C.E.R.L. Freshwater Biology Unit, Ratcliffe-on-Soar Power Station, Nottingham, LM/BIOL/006.

Antalfi, A. and Toelg, J., 1971: *Graskarpfen-Pflanzenfressende Fische.* Guenzburg, Donau Verlag, p. 207.

Antalfi, A. and Toelg, E., 1973: *Reproduction des carpes herbivores, traitement des larves et des alevins.* EIFAC/T 25: 111–121.

Antsyshkina, L.M., Kirilenko, N.S., Melnikov, G.B. and Ryabov, F.P. 1968: Dynamics of weight, nutritional state and body length of *Tilapia mossambica* Peters, reared on granular, *Chlorella*-containing feed. *Probl. Ichthyol.*, **4**, 576–580.

Anwand, K., 1963: Die Wirkung von Hypophysen- und Gonabioninjektionen auf Hechtmilchner. *Deutsche Fischerei-Zeitung*, **10**, 202–207.

Anwand, K., Mende, R., Schlumpberger, W., Hillenbrand, M. and Liebenau, H., 1976: Ergebnisse der Entwicklung und Erprobung von Trockenmischfuttermitteln fuer die industriemaeßige Aufzucht von Karpfenbrut in Warmwasseranlagen. *Zeitschr. Binnenfisch. DDR*, **23**, H.7, 194–215.

215

Atz, J.W. and   1959:   The use of pituitary hormones in fish culture. *Endeavour,* **18,**
Pickford, G.E.,   125–129.

Atz, J.W. and   1964:   *The pituitary gland and its relation to the reproduction of fishes in*
Pickford, G.E.,   *nature and in captivity.* FAO Fisheries Biology, Technical Paper
                               No. 37.

Austreng, E.,   1978:   Fat-containing Bleaching Earth as a Feed Constituent for
                               Rainbow Trout. *Aquaculture,* **15,** 333–343.

Babajan, K.,   1974:   Waemeenergetik und industriemaeßige Fischaufzucht (in
                               Russian). See: Merla, G., *Zeitschr. Binnenfischerei DDR,* **22,**
                               H.12, 380–383 (1975).

Bakos, J.,   1967:   Die Notwendigkeit der Selektionsarbeit und ihre Verfahren in
                               den Karpfenteichwirtschaften (Hungarian). *Halaszat,* **13,** (60),
                               132–133.

Bakos, J.,   1979:   *Crossbreeding Hungarian races of common carp to develop more*
                               productive hybrids. Advances in Aquaculture, FAO Technical
                               Conference on Aquaculture, Kyoto FIR: AQ/Conf/76/E74.

Bank, O.,   1967:   Karpfenzucht und Karpfenfrostung in Venezuela. *Allg.*
                               Fischerei-Zeitung, **92,** 377–378.

Bardach, J.E.,   1972:   *Aquaculture.* Wiley Interscience. New York.
Ryther, J.H. and
McLarney, W.O.,

Barnard, J.L.,   1973:   *Biological Denitrification.* S. African Branch Water Poll.
                               Control.

Beamish, F.W.H.,   1964:   Respiration of fishes with special emphasis on standard oxygen
                               consumption. *Canadian Journal of Zoology,* **42,** 177–188.

Beck, H.,   1978:   *Single Cell Proteins in Trout Diets.* Proc. World Symp. on
Gropp, J.,   Finfish Nutrition and Feed Technology, Hamburg.
Koops, H. and
Tiews, K.,

Beck, H.,   1977:   Weitere Moeglichkeiten des Fischmehl-Ersatzes im Futter fuer
Koops, H.,   Regenbogenforellen: Ersatz von Fischmehl durch Alkanhefe
Tiews, K. and   und Krillmehl. *Arch. Fisch. Wiss.,* **28,** H.1, 1–17.
Gropp, J.,

Berger, M.,   1977:   Ein neues Verfahren zur Intensivproduktion von Speisefischen.
                               Fischer und Teichwirt, **9.**

Berger, M.,   1977:   Silox-Fischzuchtanlage: Ein neues Verfahren zur Intensiv-
                               produktion von Speisefischen. *Fischer und Teichwirt,* **28,** H.9,
                               116–118.

Bishai, H.M.,   1974:   Fecundity of the mirror carp (*Cyprinus carpio L.*) at the Serow
Ishak, M.M. and   Fish Farm (Egypt). *Aquaculture,* **4,** 257–265.
Labib, W.,

Boetius, I. and   1967:   Studies in the European Eel, Anguilla anguilla (L). Exper-
Boetius, J.,   imental induction of the male sexual cycle, its relation to
                               temperature and other factors. *Meddelelser fra Danmarks*
                               Fiskeri- og Havundersøgelser, **4,** H.11, 339–405.

Bohl, M.,   1970:   Zwischenbericht ueber Aalfuetterungsversuche in Wielenbach.
                               Allg. Fischerei-Zeitung, **95,** 340–341.

Bok, A.H.,   1980:   Production and growth of carp (Cyprinus carpio) at Amalinda
                               Fish Station, eastern Cape. *S. Afr. J. Wildl. Res.,* **10,** 140–149.

Broussard, M.C. Jr. and 1976: High-Density culture of channel catfish in a recirculating system
Simco, B.A.,   *The Progressive Fish-Culturist,* **38,** 3, 138–141.

Burzev, J.A.,   1969:   Brutgewinnung von Zwischengattungshybriden Hausen x Ster-
                               let (in Russian). In: Golovimskaja, Kirpitchnikov und Kuder-
                               ski: *Genetik, Selektion und Hybridisation der Fische* (in
                               Russian). Wissenschaft, Moskau.

Buschkiel, A.,   1932:   Studien ueber das Wachstum von Fischen in den Tropen. I.
                               Udber Karpfen in Indien und ihr Wachstum. Internat. *Rev. d.*
                               ges. Hydrobiologie und Hydrographie, **27.**

Buschkiel, A.,	1933:	Teichwirtschaftliche Erfahrungen mit Karpfen in den Tropen. *Z. f. Fischerei*, **31**.
Buschkiel, A.,	1939:	Stoffwechsel im tropischen Teich, fischereibiologisch betrachtet. *Arch. f. Hydrobiologie*, Suppl. **16** (Trop. Binnengewaesser 8), 156.
Cairns, J., Heath, A.G. and Parker, B.C.,	1975:	Temperature influence on chemical toxicity to aquatic organisms. *Journal WPCF,* **47**, 2, 267–280.
Cansdale, G.,	1980:	Personal communication (editor). Seawater Supplies, Gt Chesterford, Essex, U.K.
Celikkale, M.S.,	1976:	*Ahrensburg ve Dinkelsbuehl Aynah Sazan Hatlarinin Ayni Cevre Kosullarindaki Bueyueme ve Doel Verim Oezelliklerinin Karsilastirmasi Uezerinde Arastirmalar.* Dept of Agriculture, Ankara University, Turkey.
Cellarius, O.,	1973:	Bastardierungsversuche mit Cypriniden. *Aquarien-Magazin,* **7**, 70–72.
Chaudhuri, H.,	1976:	Use of Hormones in Induced Spawning of Carps. *J. Fish. Res. Board Can.,* **33**, 4, Pt. 2, S.940–947.
Chaudhuri, H., Juario, J.V., Primavera, J.H., Samson, R. and Mateo, R.,	1978:	Observations on artificial fertilization of eggs and the embryonic and larval development of milkfish, *Chanos chanos* (Forskal). *Aquaculture* **13**, 95–113.
Chiba, K.,	1980:	*Present Status of Flow-Through and Recirculation Systems and Their Problems in Japan.* FAO EIFAC/80/Symp.: R 16, 1–16.
Colak, A. and Yamamoto, K.,	1974:	Ultrastructure of the Japanese eel spermatozoon. *Annotationnes Zoologicae Japonenses,* **47**, 1, 48–54.
Colberg, P.J. and Lingg, A.J.,	1978:	Effect of ozonation on microbial fish pathogens, ammonia, nitrate, nitrite, and BOD in simulated reuse hatchery water. *J. Fish. Res. Board Can.,* **35**, 1290–1296.
Collis, W.J. and Smitherman, R.O.,	1978:	*Production of Tilapia Hybrids with Cattle Manure or a Commercial Diet.* Symposium on the culture of exotic fishes, Fish Culture Section, American Fisheries Society.
Cruz, E.M. and Laudencia, I.L.,	1978:	Screening of feedstuffs as ingredients in the rations of Nile tilapia. (Kalikasan) *Philipp. J. Biol.,* **7** (2), 159–164.
Csavas, I. and Varadi, L.,	1980:	*Design and Operation of a Large-Scale Experimental Recycling System Heated with Geothermal Energy at the Fish Culture Research Institute, Szarvas, Hungary.* FAO EIFAC/80/Symp.: E 16, 1–15.
Dadzie, S.,	1970:	Preliminary report on induced spawning of *Tilapia aurea. Bamidgeh,* **22**, 1.
Deelder, C.C.,	1970:	*Synopsis of biological data on the eel* Anguilla anguilla *(Linnaeus 1758).* FAO Fish Synopsis No. 80, Rome.
De Silva, S.S., and Weerakoon, D.E.M.,	1981:	Growth, Food Intake and Evacuation Rates of Grass Carp, *Ctenopharyngodon Idella* Fry. *Aquaculture,* **25**, 67–76.
Donaldson, E.M.,	1977:	*Bibliography of fish reproduction 1963–1974* (3 parts). Fisheries and Marine Service Technical Report 732, West Vancouver, British Columbia.
Drew, K.M.,	1949:	See: Bardach, Ryther and McLarney, p. 792.
Drew, K.M.,	1956:	Reproduction in the Bangiophycidae. *Bot. Rev.,* **22**, 553–611.
Eckhardt, O.,	1981:	*Untersuchungen zum Protein- und Energiebedarf junger wachsender Spiegelkarpfen* (Cyrinus carpio L.). Diss., Landwirtsch. Fak., Univ. Gottingen.
Fijan, N.,	1973:	*Induced Spawning, Larval Rearing and Nursery Operations — Silurus glanis* EIFAC/T25: 130–135.

Fischer, G. and   1971:   Ein biologisch wirksames Peptid aus der Haut von Neunaugen.
Albert, W.,   *Naturwissenschaften,* **58**, 363.

Fischer, Z.,   1973:   The elements of energy balance in grass carp (Ctenopharyn-
                                  godon idella Val.) Part IV. Consumption rate of grass carp fed
                                  on different type of food. *Pol. Arch. Hydrobiol.,* **20**, 309–318.

Fishelson, K.,   1962:   Hybrids of two species of fishes of the genus *Tilapia. Fisher-
                                  man's Bull. Haifa,* **4**, 14–19.

Fontaine, M.,   1964:   Sur la maturation des organes génitaux de l'Anguille femelle
Bertrand, E.,   (*Anguilla anguilla* L.) et l'émission spontanée des oeufs en
Lopez, E. and   aquarium. *C.R. Acad. Sci. (Paris),* **259**, 2907–2910.
Callamand, O.,

Forster, J.R.M.,   1974:   Studies on nitrification in marine biological filters. *Aquaculture,*
                                  4, 387–397.

Frenzel, E. and   1982:   Untersuchungen ueber den Mineralstoff — Bedarf von Regen-
Pfeffer, E.,   bogenforellen (Salmo gairdneri, R.) *Arch. Tierernaehrung,* **32**,
                                  H.1, 1–8.

Gerbilski, H.L.,   1941:   *Die Methoden der Hypophyseninjektionen und ihre Bedeutung
                                  fuer die Regeneration der Fischbestaende,* (in Russian). Isd.
                                  LGU.

Godfriaux, B.L. and   1981:   *Evolution of the Design and Operation of a Freshwater Waste
Stolpe, N.E.,   Heat Aquaculture Facility at an Electric Generating Plant,* Bio-
                                  Engineering Symposium for Fish Culture. (FCS Publ. 1):
                                  259–265.

Goetsch, W.,   1924:   Lebensraum und Koerpergroeße. *Biol. Zenralblatt,* **44**.

Goetting, K.-J.,   1971:   Wie fruchtbar sind die Meeresfische? *Naturwiss. Rundschau,*
                                  24, 100–101.

Golovinskaya, K.A.,   1968:   Genetics and selection of fish and artificial gynogenesis of the
                                  carp (*Cyprinus carpio*). *FAO Fish. Rep.,* (44) **4**, 251–22.

Goslar, H.G.,   1969:   Die Wirkung eines Milzdialysates auf das Enzymmuster des
Grigoriadis, P. and   infantilen Meerschweinchen — und Rattenhodens am normalen
Jaeger, K.H.,   und hypophysektomierten Tier. *Arzneimittel-Forschung,* **19**,
                                  1249–1253.

Grigoriadis, P.,   1969:   Vergleichende enzymhistochemische Untersuchungen zum
Goslar, H.G. and   Verhalten des infantilen Meerschweinchenhodens nach Milzex-
Jaeger, K.H.,   trakt — und Gonadotropinapplikation. *Verh. d. Anat. Gesell.,*
                                  Ergaenzungsheft zu Band 125.

Gross, R.,   1978:   Nutritional tests with green alga *Scenedesmus* with health and
Gross, U.,   malnourished persons. *Arch. Hydrobiol. Beih.* **11**, S.161–173.
Ramirez, A.,
Cuadra, K.,
Collazos, C. and
Feldheim, W.,

Guenther, K.D.,   1972:   Wachstum und Mineralumsatz. In: *Handbuch der Tierernaeh-
                                  rung II* (Hrsg. W. Lenkeit und Breirem, K.) Teil V, Kap. 4,
                                  Paul Parey, Hamburg und Berlin.

Guerbrero, R.D.,   1975:   Sex-reversal of *Tilapia. FAO Aquaculture Bull.,* **7**, 3–4, 7.

Guerrero, R.D.,   1975:   Use of androgens for the production of all-male *Tilapia aurea*
                                  (Steindachner), *Trans. Am. Fish. Soc.* **104**, 342–348.

Guerrero, R.D.,   1979:   *Culture of male* Tilapia mossambica *produced through artificial
                                  sex reversal.* Advances in Aquaculture, FAO Technical Confer-
                                  ence on Aquaculture, Kyoto 1976. Fishing News Books.

Gupta, S.,   1975:   The development of carp gonads in warm water aquaria. *J. Fish
                                  Biol.,* **7**, 775–782.

Gupta, S. and   1976:   Abnormalities of the gonads of carp. *J. Fish Biol.,* **9**, 75–77.
Meske, C.,

Halver, J.E.,   1978:   *Vitamin Requirements of Finfish.* Proc. World Symp. on Finfish
                                  Nutrition and Feed Technology, Hamburg.

Hambrey, J.,	1980:	*The Importance of Feeding, Growth and Metabolism in Consideration of the Economics of Warm Water Fish Culture Using Waste Heat.* FAO EIFAC/80/Symp., E 74: 1–17.
Hambrey, J.,	1981:	*Technical and economic consequences of using waste heat for aquaculture.* Proc. Int. Sem. Energy Conservation and Use of Renewable Energies in the Bio-industries. Pergamon Press.
Hamm, A.,	1964:	Die Hitzeresistenz des Karpfens mit und ohne Anpassung an hoehere Temperaturen. *Allg. Fischerei-Zeitung*, **89**, 488–489.
Haug, R.T. and McCarty, L.P.	1972:	Nitrification with submerged filters. *Journal WPCF*, **44**, No. 11, 2086–2102.
Hepher, B., Sandbank, E. and Shelef, G.,	1978:	*Alternative Protein Sources for Warmwater Fish Diets.* Proc. World Symp. on Finfish Nutrition and Feed Technology, Hamburg.
Hickling, C.F.,	1963:	The Cultivation of Tilapia. *Scient. Am.*, **208**, 5, 143–150.
Hilge, V.,	1976:	Ergebnisse zur kuenstlich eingeleiteten Geschlechtsreifung bei Aalen. *Arbeiten des deutschen Fischerei-Verbandes*, H. **19**, 117–124.
Hilge, V.,	1978:	*Preliminary Results with Krill Meal and Fish Meal in Diets for Channel Catfish (Ictalurus punctatus Raff.)* Proc. World Symp. on Finfish Nutrition and Fishfeed Technology, Hamburg.
Hilge, V.,	1980:	*Rearing of Channel Catfish (Ictalurus punctatus Raff.) in a closed warm water system.* FAO EIFAC/80/Symp., E5: 1–11.
Hilge, V. and Conrad, C.,	1975:	Zum Laichvermoegen eines hermaphroditen Karpfens (*Cyprinus carpio*). *Arch. Fisch Wiss.*, **26**, H.1, 49–51.
Hirayama, K.,	1974:	Water control by filtration in closed culture systems. *Aquaculture*, **4**, 369–385.
Hoar, W.S.,	1969:	Hoar, W.S. and Randall, D.J. Reproduction. In: *Fish Physiology*, Vol. III: 1–72. Academic Press.
Hoffbauer, C.,	1902:	Ueber den Einfluß des Wasservolumens auf das Wachstum der Fische. *Allg. Fischerei-Zeitung N.F.*, **17**, 103.
Holtz, W., Bueyuekhatipoglu, S., Stoss, J., Oldigs, B. and Langholz, H.J.,	1979:	*Preservation of Trout Spermatozoa for Varying Periods.* Advances in Aquaculture. FAO Technical Conference on Aquaculture, Kyoto 1976. Fishing News Books.
Honma, J.,	1980:	*Aquaculture in Japan.* Japan FAO Assoc., Tokyo.
Horvath, L.,	1977:	Improvement of the method for propagation, larval and postlarval rearing of the wels (Silurus glanis L.). *Aquaculture* **10**, S.161–167.
Horvath, L.,	1978:	*The Rearing of Warmwater Fish Larvae.* Proc. World Symp. on Finfish Nutrition and Feed Technology, Hamburg.
Horvath, L.,	1979:	*Indoor and pond rearing technology for fry and fingerlings of wels/Silurus glanis L.* EIFAC/T 35, Suppl. 1, 85–93.
Howell, B.R.,	1981:	*The use of heat in fish farming.* Proc. 1st Int. Sem. Energy Conservation and Use of Renewable Energies in the Bio-industries. Pergamon Press.
Huisman, E.A.,	1976:	*Sauerstoff als Umweltfaktor fuer die Fischproduktion* The influence of environmental factors upon the health of fishes. Fischer, Stuttgart.
Huisman, E.A.,	1978:	*The Culture of Grass Carp (Ctenopharyngodon idella Val.) under artificial Conditions. An experience paper.* EIFAC/78/Symp:E/48.
Huisman, E.A.,	1981:	*Integration of Hatchery, Cage and Pond Culture of Common Carp (Cyprinus carpio L.) and Grass Carp (Ctenopharyngodon idella Val.) in the Netherlands.* Bio-Engineering Symposium for Fish Culture (FCS Publ. 1): 266–273.
Ihering, G.,	1935:	Die Wirkung von Hypophyseninjektionen auf den Laichakt von Fischen. *Zool. Anz.* **3**.

Jauncey, K., 1982: The effects of varying dietary protein level on the growth, food conversion, protein utilization and body composition of juvenile tilapias (Sarotherodon mossambicus). *Aquaculture, 27*, 43–54.

Jirasek, J., 1976: Moeglichkeiten der Anfangsfuetterung der Karpfenbrut mit Ersatzfuttermitteln. *Zivocisna vyroba* 21 (12) S.871–879.

Johnson, W.K. and Schroepfer, G.J., 1964: Nitrogen removal by nitrification and denitrification. *J. Water Poll. Contr. Fed., 36*, 1015–1036.

Koendler, R., 1971: Ueber Vorkommen und Haeufigkeit der Zwischenmuskelgraeten bei Fischen aus Sueß- und Salzwasser. *Fischwirt, 21*, 97–111.

Kausch, H., 1975: Der Einfluß der Wassertemperatur auf die Inkubationszeit bei der Hypophysierung von Karpfen (*Cyprinus carpio* L.) unter Freilandbedingungen. *Arch. Hydrobiol.*, Suppl. *47*, 3, 413–422.

Kausch, H. and Ballion-Cusmano, M.R., 1976: Koerperzusammensetzung, Wachstum und Nahrungsnutzung bei jungen Karpfen unter Intensivhaltungsbedingungen. *Arch. Hydrobiol.*, Suppl. *48*, H.2, 141–180.

Keiz, G., 1963: Vergleich der Wachstumsleistung importierter Satzkarpfen aus Israel mit Karpfen Wielenbacher Herkunft. *Allg. Fischerei-Zietung, 88*, 551.

Kincannon, D.F. and Gaudy, A.F., 1968: Response of biological waste treatment systems to changes in salt concentrations. *Biotechnol. Bioeng., 10* (4), 469–483.

Kirk, R.G., 1972: A review of recent developments in *Tilapia* culture, with special reference to fish farming in the heated effluents of power stations. *Aquaculture* 1 1, 45–60.

Kirpichnikov, V.S., 1971: *Methods of fish selection. 1. Aims of selection and methods of artificial selection.* U.N. Dev. Progr. No. Ta 2926, 202–216, FAO.

Kirpichnikov, V.S., 1981: *Genetic Bases of Fish Selection.* Springer-Verlag.

Kirpichnikov, V.S., Factorovich, K.A., Ilyasov, Yu.I. and Shart, L.A., 1979: *Selection of common carp* (Cyprinus carpio) *for resistance to dropsy.* FAO Technical Conference on Aquaculture, Kyoto, FIR: AQ/Conf/76/E.63.

Klinger, H., 1977: Jahresbericht der Bundesforschungsanstalt fuer Fischerei, Hamburg, F32 and unpublished data.

Klinger, H. and Meske, C., 1978: unpublished. (personal comm.)

Knoesche, R., 1969: Stoerzucht- eine Methode zur Bereicherung des Fischangebotes und zur Steigerung der Tentabilitaet der Binnenfischerei. *Deutsche Fischerei-Zeitung, 16*, 136–141.

Knoesche, R., 1969: *Untersuchungen ueber die technischen Voraussetzungen zur Schaffung optimaler Umweltbedingungen bei der industriemaeßigen Fischproduktion unter besonderer Beruecksichtigung geschlossener Kreislaeufe.* Promotion, Inst. Binnenfisch., Berlin-Friedrichshagen, DDR.

Knoesche, R., 1974: Der Waermehaushalt von Fischzuchtanlagen. *Zeitschr. Binnenfischerei DDR, 21*, H.3, 77–84.

Knoesche, R. and Tscheu, E., 1974: Das Belebtschlammverfahren zur Reinigung des Abwassers aus Fischzuchtanlagen. *Z. Binnenfischerei DDR, 1*, 16–23.

Konikoff, M. and Lewis, W., 1974: Variation in weight of cage reared channel catfish. *Prog. Fish. Cult., 36*, 3, 128–144.

Koops, H., 1965: Fuetterung von Aalen in Teichen. *Arch. Fischereiwiss., 16*, 33–38.

Koops, H., 1966: Ein Fuetterungsversuch mit Satzaalen in der Flußteichwirtschaft Mueden an der Mosel. *Arch. Fischereiwiss., 17*, 36–44.

Koops, H., 1967: Die Aalproduktion in Japan. *Arch. Fischereiwiss., 17*, 43–50.

Koops, H., 1973: Hohe krankheitsbedingte Sterblichkeit behindert Versuche zur Entwicklung einer Aalteichwirtschaft. *Infn Fischw., 20* (3), 80–81.

Koops, H. and
Kuhlmann, H.,   1978:   Zur Entwicklung von Aalmastfuttern *Infm. Fischw.*, **25** (1).

Koops, H. and
Kuhlmann, H.,   1980:   *Eel farming in the thermal effluent of a conventional power station at Emden harbour.* FAO EIFAC/80/Symp., E 4: 1–16.

Koske, P.H.,
Lenz, J.,
Nellen, W. and
Zeitzschel, B.,   1973:   *Die Produktion mariner Organismen unter natuerlichen Verhaeltnissen und in Kulturen.* Studie fuer das Bundesministerium fuer Bildung und Wissenschaft, Kiel.

Kossmann, H.,   1970:   Versuch zur Erhoehung der Zuwachsleistung von Karpfen in Teichwirtschaften durch gezielte Bruterzeugung im Warmwasser. 1. *Mitteilung. Fischwirt*, **20**, 255–263.

Kossmann, H.,   1971:   Hermaphroditismus und Autogamie beim Karpfen. *Naturwissenschaften*, **58**, H.6, 328–329.

Kossmann, H.,   1972:   Untersuchungen ueber die genetische Varianz der Zwischenmuskelgraeten des Karpfens. *Theoretical and Applied Genetics*, **42**, 3, 130–135.

Kossmann, H. and
Szablewski, W.,   1971:   Versuch zur Erhoehung der Zuwachsleistung von Karpfen in Teichwirtschaften durch gezielte Bruterzeugung im Warmwasser. 2. *Mitteilung. Fischwirt*, **21**, 49–53.

Kraut, H. and
Meffert, M.-E.,   1966:   *Ueber unsterile Großkulturen von* Scenedesmus obliquus. Westdeutscher Verlag.

Krupauer, V.,   1963:   Vliv velikosti životního prostředí na ŕust kapra. *Ceskoslov. Rybarstvi*, **85**.

Kuhlmann, H.,   1975:   Der Einfluss von Temperatur, Futter, Groesse und Herkunft auf die sexuelle Differenzierung von Glasaalen (Anguilla anguilla). *Helgolander wiss. Meeresunters.*, **27**, 129–155.

Kuhlmann, H.,   1979:   Neue Haltungstechnik zur Anfuetterung von Glasaalen. *Infn Fischw.*, **26** (5), 144–146.

Ladiges, W. and
Vogt, D.,   1965:   *Die Süßwasserfische Europas.* Parey, Hamburg, Berlin.

Lall, S.P.,   1978:   *Minerals in Finfish Nutrition.* Proc. World Symp. on Finfish Nutrition and Feed Technology, Hamburg.

Langhans, V.,   1928:   Der "Raumfaktor" in der praktischen Teichwirtschaft. *Nachrichtenblatt fuer Fischzucht und Fischerei*, **1**, 7.

Langhans, V. and
Schreiter, T.,   1928:   Die Raumfaktorversuche an der staatlichen Forschungsstation fuer Fischzucht und Hydrobiologie in Hirschberg. *Nachrichtenblatt fuer Fischzucht und Fischerei*, **1**.

Lechler, H.,   1934:   Ueber das Wachstum der Fische. 1. Teil: Die Wirkungskraefte des Wachstums. *Z. f. Fischerei*, **32**, 281.

Liao, P.B.,   1980:   *Treatment Units Used in Recirculation Systems for Intensive Aquaculture.* FAO EIFAC/80/Symp., R 6: 1–23.

Lieder, U.,   1961:   Untersuchungsergebnisse ueber die Gratenzahl bei 17 Süßwasser-Fischarten. *Z. f. Fischerei*, **10**, 329–350.

Lieder, U. and
Jaehnichen, H.,   1975:   Versuche ueber Moglichkeiten zur Anfuetterung frischgeschluepfter Amurkarpfenbrut (Ctenopharyngodon idella) mit einem speziellen Trockenfuttermittel. *Zeitschr. Binnenfisch. DDR*, **22**, H.4, 118–120.

Loerz, R.,
Lukowicz, M.v. and
Jahn, F.,   1977:   Die Zusammensetzung von Krillmehl. *Fischwirt*, **27**, 9, 50–51.

Lueling, K.H.,   1965:   Jungtiere des groeßten Sueßwasserfisches der Erde zum ersten Mal lebend in Europa. *Der Zoologische Garten (NF)*, **31**, 295–303.

Lukowicz, M.v.,   1976:   Versuche zur Anfuetterung von Karpfenbrut. *Arbeiten des Deutschen Fischerei-Verbandes*, H. **19**, 107–116.

Lukowicz, M.v. and
Rutkowski, F.,   1976:   Anfuetterung von Karpfenbrut. *Fischer und Teichwirt*, **27**, H.6, 68–69.

Luther, G.,   1970:   Anwendungsmoeglichkeiten neuer wissenschaftlicher Erkennt-

Meske, C.,   1968:   Hypophysierung von Aquarienkarpfen und kuenstliche Lai-
Woynarivich, E.,   cherbruetung als Methoden zur Zuechtung neuer Karpfen-
Kausch, H.,   rassen. *Theoretical and Applied Genetics,* **38**, 47–51.
Luhr, B. and
Szablewski, W.,

Meyer-Waarden, P.F.,   1965:   Die wundersame Lebensgeschichte des Aales. In: Keune: *Der
                                      Aal.* Hans A. Keune Verlag, Hamburg.

Meyer-Waarden, P.F.,   1967:   Maßnahmen zur Intensivierung der Aalwirtschaft in der Bun-
                                      desrepublik. In: Die Aalwirtschaft in der Bundesrepublik
                                      Deutschland. *Arch. Fischereiwiss.,* **18**, 2. Beiheft, 497–567.

Miaczynski, T. and   1961:   Wzrost karpi poczatkowo przetrzymanych w akwariach. *Acta
Rudzinski, E.,   hydrobiologia* (Krakau), **3**, 165–174.

Mironova, N.V.,   1969:   Comparison of growth of Tilapias (*Tilapia mossambica* Peters)
                                      when fed *Chlorella* and other foodstuffs. *NASA Tech. Transl.
                                      TTF,* **529**, 478–484.

Mironova, N.V.,   1974:   The energy balance of *Tilapia mossambica. J. Ichthyol.,* **3**,
                                      431–438.

Mitterstiller, J. and   1961:   Foerderung des Ablaichens beim Karpfen durch Hormonprae-
Hamor, T.,   parate. *Deutsche Fischerei-Zeitung,* **8**, 117–118.

Moav, R.,   1979:   *Genetic Improvement in Aquaculture Industry.* Advances in
                                      Aquaculture, FAO Technical Conference on Aquaculture,
                                      Kyoto 1976, Fishing News Books.

Moav, R.,   1975:   Variability of intermuscular bones, vertebrae, ribs, dorsal fin
Finkel, A. and   rays and Skeletal disorders in the common carp. *Theoretical and
Wohlfarth, G.W.,   Applied Genetics,* **46**, 33–43.

Moav, R. and   1966:   Genetic improvement of yield in carp. *FAO Fish. Rep.,* **44**,
Wohlfarth, G.W.,   12–29.

Moav, R. and   1967:   *Breeding schemes for the genetic improvement of edible fish.*
Wohlfarth, G.W.,   Prog. Rep. Israel, U.S. Fish Wildl. Serv., 1964–1965.

Moav, R. and   1976:   Two-way selection for growth rate in the common carp
Wohlfarth, G.W.,   (*Cyprinus carpio* L.). *Genetics,* **82**, 83–101.

Moav, R.,   1960:   Genetic improvement of carp II: marking fish by branding.
Wohlfarth, G.W. and   *Bamidgeh,* **12**, 49–53.
Lahman, M.,

Molnar, G. and   1962:   Experimente mit Welsen (Silurus glanis L.) zur Feststellung des
Toelg, I.,   Zusammenhanges der Temperatur und der Zeitdauer der
                                      Magenverdauung. *Annal. Biol. Tihany,* **29**, 107–115.

Mudrack, K.,   1970:   *Untersuchungen ueber die Anwendung der mikrobiellen Denitri-
                                      fikation zur biologischen Reinigung von Industrieabwasser.*
                                      Veroeffentlichungen des Institutes fuer Siedlungswasser-
                                      wirtschaft der TU Hannover, H. 36.

Mudrack, K.,   1976:   Untersuchungen ueber den Einfluss der Salzkonzentration auf
                                      den biologischen Abbau von Wasserverunreinigungen. *Wasser-
                                      und Abwasserforschung,* **9**, 6, 179–182.

Mueller, H.,   1964:   Wachstum, Fuetterung, Markierung und Fang von Aalen in
                                      kleinen Teichen. *Z. f. Fischerei,* **12**, 189–200.

Mueller, H.,   1967:   Ueber den Stand der Aalintensivhaltung. *Deutsche Fischerei-
                                      Zeitung,* **14**, 1–7.

Mulbarger, M.C.,   1971:   Nitrification and denitrification in activated sludge systems. *J.
                                      Water Poll. Contr. Fed.,* **43**, 2059–2070.

Naegel, L.,   1976:   Untersuchungen zur Intensivhaltung von Fischen im Warm-
Meske, C. and   wasserkreislauf. *Arch. Fisch Wiss.,* **27**, H.1, 9–23.
Mudrack, K.,

Nagy, A.,   1978:   Investigation on carp, *Cyprinus carpio* L. gynogenensis. *J. Fish
Rajki, A.,   Biol.,* **13**, 215–224.
Horvath, L. and
Csanyi, V.,

Nakaruma, N. and Kasahara, S., 1955: *see*: Wohlfarth, G.W., 1977: *Bamigdeh,* **29**, 2, 35–56.

Nakaruma, N. and Kasahara, S., 1956: *see*: Wohlfarth, G.W., 1977: *Bamigdeh,* **29**, 2, 35–56.

Nakaruma, N. and Kasashara, S., 1957: *see*: Wohlfarth, G.W., 1977: *Bamigdeh,* **29**, 2, 35–56.

Nakaruma, N. and Kasashara, S., 1961: *see*: Wohlfarth, G.W., 1977: *Bamigdeh,* **29**, 2, 35–56.

Neal, R.A. and Mock, C.R., 1979: *A model closed system for aquaculture incorporating the recycling of wastes.* FAO Technical Conference on Aquaculture, Kyoto. FIR: AQ/Conf/76/E.22.

Nemoto, K., 1980: *Eel Culture in a Recirculation and Filtration System Utilizing Heated Fresh Water Effluents.* FAO EIFAC/80/Symp.: E 49, 1–19.

Neudecker, T., 1976: Die Embryonalentwicklung des Karpfens (*Cyprinus carpio* L.). *Arch. Fisch Wiss.,* **27**, H.1, 25–35.

Neudecker, T. 1978: Erste Erfolge bei der kuenstlichen Vermehrung der Pazifischen Auster (Crassostrea gigas) in der Bundesrepublik Deutschland. *Inf.f.d. Fischwertschaft* **25** (2), 48–49.

Nikoljukin, N.J., 1966: Einige Fragen der Zytogenetik, Hybridisation und Systematik bei Acipenseriden (in Russian). *Genetika Moskva,* **5**, 25–27.

Nose, T., 1971: Spawning of eel in a small aquarium. *Piscicoltura e Ittiopatologia,* **6**, 25–26.

Nose, T., 1978: *Diet Compositions and Feeding Techniques in Fish Culture with Complete Diets.* Proc. World Symp. on Finfish Nutrition and Feed Technology, Hamburg.

Nose, T. and Arai, S., 1979: *Recent Advances in Studies on Mineral Nutrition of Fish in Japan.* Advances in Aquaculture. FAO Technical Conference on Aquaculture, Kyoto, 1976. Fishing News Books.

Nuorteva, P., 1971: Methylquecksilber in den Nahrungsketten der Natur. *Naturw. Rundschau,* **24**, 233–243.

Olivereau, M., 1961: *Maturation sexuelle de l'anguille male en eau douce.* Compt. Rend. Acad. Sci. Paris 252 (23): 3660–3662.

Pabst, W., 1975: Die Massenkultur von Mikroalgen — Stand der Technik und Bewertung der Produkte. *Kraftfutter,* **58**, H.2.

Pearson, E.A. and Frangipane, E. De Fraga (Eds.), 1975: *Marine Pollution and Marine Waste Disposal.* Pergamon Press, Oxford, 487 pp.

Peters, G., Delventhal, H. and Klinger, H., 1980: Physiological and morphological effects of social stress in the eel. (*Anguilla anguilla*). *Arch. Fisch Wiss.,* **30**, 2/3, 157–180.

Pfeffer, E., 1978: Ueber die Verteilung von mineralischen Mengenelementen im Koerper von Forellen und Karpfen. *Zeitschrift fuer Tierphysiologie, Tierernaehrung und Futtermittelkunde,* **40**, H.3, 159–164.

Pfeffer, E. and Becker, K., 1977: Untersuchungen an Regenbogenforellen ueber den Futterwert verschiedener Handelsfutter und ueber den weitgehenden Ersatz von Fischmehl durch Krillmehl im Futter. *Arch. Fisch Wiss.,* **28**, 1, 19–29.

Pfeffer, E., Matthiesen, J., Potthast, V. and Meske, C., 1977: Untersuchungen an Karpfen ueber die Zusammensetzung des Zuwachses bei unterschiedlichen Protein-Gehalten im Futter. *Fortschritte in der Tierphysiologie und Tierernaehrung,* H.8, 19–31.

Pfeffer, E. and Meske, C., 1978: Untersuchungen ueber Casein und Krillmehl als einzige Proteinquelle im Alleinfutter fuer Karpfen. *Zeitschrift fuer Tierphysiologie, Tierernaehrung u. Futtermittelkunde,* **40**, H.2, S.74–91.

Pfeffer, E.,
Pieper, A.,
Matthiesen, J. and
Meske, C.,
1977:   Untersuchungen zum Hungerumsatz von Karpfen. *Fortschritte in der Tierphysiologie und Tierernaehrung*, H.8, 8–18.

Pfeffer, E. and
Potthast, V.,
1977:   Untersuchungen ueber den Ansatz von Energie, Protein und mineralischen Mengenelementen bei wachsenden Regenbogenforellen. *Fortschritte in der Tierphysiologie und Tierernaehrung*, H.8, 32–55.

Piekarski, G.,
1939:   Der Einfluß der Raumgroeße auf den Organismus. *Reichs-Gesundheitsblatt*, H.37, 759–762.

Pillay, T.V.R.,
1976:   *The state of aquaculture 1975.* FAO Technical Conference on Aquaculture, Kyoto, FIR:AQ/Conf/76/R.36.

Pohlhausen, H.,
1978:   *Lachse in Seen, Teichen, Fluessen und Baechen.* Paul Parey, Hamburg and Berlin.

Precht, H.,
Christopherson, J.,
Hensel, H. and
Larcher, W.,
1973:   *Temperature and life.* Springer Verlag Berlin, Heidelberg, New York, 770 pp.

Price, K.S.,
1978:   Advances in closed (recirculated) system mariculture *Rev. Biol. Trop.*, **26** (Suppl. 1): 23–43.

Probst, E.,
1943:   Kreuzungsversuche bei Karpfen und ihre Bedeutung fuer die Leistungszucht. *Fischerei Zeitung*, **47**, H.1/2.

Pruginin, Y.,
Rothbard, S.,
Wohlfarth, G.,
Halevy, A.,
Moav, R. and
Hulata, G.,
1975:   All-male broods of *Tilapia nilotica x T. aurea* hybrids. *Aquaculture*, **6**, 11–21.

Purdom, C.E.,
1969:   Radiation-induced gynogenesis and androgenesis in fish. *HEREDITY, London*, **24**, 431–444.

Purdom, C.E.,
1972:   *Genetics and fish farming.* Laboratory leaflet (new series) No. 25 (1972).

Purdom, C.E.,
1972:   Induced polyploidy in Plaice (*Pleuronectes platessa*) and its hybrid with the Flounder (*Platichthys flesus*) *Heredity*, **29**, 11–24.

Purdom, C.E.,
1982:   *Fish cultivation in heated effluents.* Proc. 2nd Int. Sem. Energy Conservation and Use of Renewable Energies in the Bio-industries, London. Pergamon Press.

Purdom, C.E. and
Lincoln, R.F.,
1973:   Chromosome manipulation in fish. In: *Genetics and mutagenesis of fish.* Edited by J.H. Schroeder, Springer-Verlag, Berlin.

Rajamani, M. and
Job, S.V.,
1976:   Food utilization by *Tilapia mossambica* (Peters): Funktion of size. *Hydrobiologica*, **50**, 71–74.

Rajbanshi, K.G.,
1966:   Fuetterungsversuche mit Trockenfutter bei Karpfen. *Fischwirt*, **16**, 99–102.

Ray, L.,
1981:   *Channel Catfish Production in Geothermal Water.* Bio-Engineering Symposium for Fish Culture. (FCS Publ. 1): 192–195.

Reichenbach-Klinke, H.-H.,
1971:   Fisch und Umweltvergiftung. *Umschau*, **71**, 564.

Reichenbach-Klinke, H.H. (Ed.),
1976:   *The Influence of Environmental Factors on the Health of Fishes.* Gustav Fischer Verlag, Stuttgart, New York, 134 pp.

Reichenbach-Klinke, H.H. (Ed.),
1977:   *Values upon Toxic Residues in Fishes and Fishwaters.* Gustav Fischer Verlag, Stuttgart, New York. 28 pp.

Reichenbach-Klinke, H.-H. and
Reichenbach-Klinke, K.-E.,
1970:   Enzymuntersuchungen an Fischen. II. Trypsin- und α-Amylase-Inhibitoren. *Arch. Fischereiwiss.*, **21**, 67–72.

Reimann, K.,
1968:   Der Abbau der organischen Substanz im Meerwasser. *Wasser- und Abwasser-Forschung*, **1**, 142–148.

Reimers, J. and
Meske, C.,
1977:   Der Einfluß eines fischmehlfreien Trockenfutters auf die Koerperzusammensetzung von Karpfen. *Fortschritte in der Tierphysiologie und Tierernaehrung*, H.8, 82–90.

Rheinheimer, G., 1966: *Einige Beobachtungen ueber den Einfluß von Ostseewasser auf limnische Bakterienpopulationen.* Veroeff. Inst. Meeresforsch. Bremerhaven Sonderband II, 237–243.

Richmond, A. and Preiss, K., 1980: The Biotechnology of Algaculture. *Interdisciplinary Science Reviews,* 5, 1.

Risa, S. and Skjervold, H., 1975: Water re-use system for smolt production. *Aquaculture,* 6, H2, 191–195.

Rogers, A. and Cane, A., 1981: *The Use of Waste Heat in Fish Farming.* Proc. Int. Sem. Energy Conservation and Use of Renewable Energies in the Bio-industries. Pergamon Press.

Rose, S. and Rose, F.C., 1965: The control of growth and reproduction in freshwater organisms by specific products. *Mitt. Internat. Verein. Limnol.* 13, 21–35.

Rosenthal, H. and Westernhagen, H.v., 1976: Fischaufzucht im Seewasserkreislauf unter Kombination biologischer und chemischer Aufbereitungsverfahren (Ozonisierung). *Arbeiten des Deutschen Fischerei-Verbandes,* H.19, 208–218.

Rueffer, H., 1964: Nitrifikation und Denitrifikation bei der Abwasserreinigung. *Vom Wass.,* 31, 134–152.

Rumohr, H., 1975: *Der Einfluß von Temperatur und Salinitaet auf das Wachstum und die Geschlechtsreife von nutzbaren Knochenfischen.* Berichte Institut fuer Meereskunde, Universitaet Kiel.

Saeki, A., 1965: Studies on fish culture in filtered closed-circulating aquaria. II. On the carp culture experiments in the systems *Bull. Jap. Soc. Sci. Fish.,* 31 (11), 916–923.

Sager, H., 1963: Die Gestalt des Goldfisches *Carassius carassius auratus* (L.) und ihre Modifikabilitaet. *Z. wiss. Zool. Abt.,* A 168, 321.

Sanchez Romero, J., 1961: *El Paiche — aspectos de su historia natural, ecologia y aprovechamiento.* Serv. Pesqueria, Min. Agricultura, Lima, 1–48.

Sarig, S. and Bejerano, I. 1980: Bacterial stress caused infection of silver carp and Sarotherodon nureus in fish ponds and their control. In: *Fish diseases* Third COPRAQ Session, Ahne, E. (ed). Springer-Verlag, Berlin, 224 pp.

Schaeperclaus, W., 1961: *Lehrbuch der Teichwirtschaft 2.* Auflage, Springer Verlag, Berlin and Hamburg.

Schaller, F. and Dorn, E., 1973: Atemmechanismus und Kreislauf des Amazonasfisches Pirarucu (*Arapaima gigas*; Pisces, Osteoglossiden). *Naturwissenschaften,* 60, 303.

Schlesner, H. and Rheinheimer, G., 1974: Auswirkungen einer Ozonisierungsanlage auf den Bakteriengehalt des Wassers eines Schauaquariums. *Kieler Meeresforschung,* XXX, H.2, 117–129.

Schlotfeldt, H.-J., 1971: Neue Moeglichkeiten zur wirtschaftlichen Aufzucht von "Teichstoeren". *Fischwirt,* 21, 9–10.

Schmeing-Engberding, F., 1953: Vorzugstemperaturen einiger Knochenfische und ihre physiologische Bedeutung. *Z. f. Fischerei,* 2, N.F., 125–155.

Schroder, J.H., 1973: *Genetics and Mutagenesis of Fish.* Springer-Verlag, Berlin, Heidelberg, New York.

Schulze-Wiehenbrauck, H., 1977: *Laborversuche ueber den Einfluß von Besatzdichte und Wasserbelastung auf* Tilapia zillii *und* Cyprinus carpio *Pisces.* Dissertation, Universitaet Kiel.

Seidlitz, U., 1975: Die industriemaeßige K_o- und K_v-Produktion in Warmwasseranlagen als Voraussetzung zur Steigerung der Ertraege zur Verbesserung der Kondition der Satzkarpfen im 1. Lebensjahr. *Zeitschr. Binnenfisch. DDR,* 22, H.2, 47–57.

Seiler, R., 1938: Die Fuetterung des Karpfens, beurteilt nach Aquarienversuchen. *Internat. Revue de Ges. Hydrobiol. und Hydrographie,* 36, 1.

Sengbusch, R.v., 1963: Fische "ohne Graeten". *Zuechter,* 33, 284–286.

Sengbusch, R.v.,   1967a: Entwicklung der neuen Haltungsmethode aus der Sicht des Zuechters. In: *Vortragsveranstaltung ueber neue Methoden der Fischzuechtung und -haltung am 15. Februar 1967*. Selbstverlag Max-Planck-Institut fuer Kulturpflanzenzuechtung, Hamburg.

Sengbusch, R.v.,   1967b: Eine Schnellbestimmungsmethode der Zwischenmuskelgraeten bei Karpfen zur Auslese von "graetenfreien" Mutanten (mit Roentgen-Fersehkamera und Bildschirmgeraet). *Zuechter*, **37**, 275–276.

Sengbusch, R.v. and   1967:   Auf dem Wege zum graetenlosen Karpfen. *Zuechter*, **37**, Meske, C.,   271–274.

Sengbusch, R.v.,   1965:   Beschleunigtes Wachstum von Karpfen in Aquarien mit Hilfe Meske, C. and   biologischer Wasserklaerung. *Experientia* **21**, 614. Szablewski, W.,

Sengbusch, R.v.,   1967:   Gewichtszunahme von Karpfen in Kleinstbehaeltern, zugleich Meske, C.,   ein Beitrag zur Aufklaerung des Raumfaktors. *Z. f. Fischerei,* Szablewski, W. and   **15**, N.F., 45–60. Luehr, B.,

Sengbusch, R.v.,   1974:   Domestikationsversuche von Fischarten aus aequatornahen Sengbusch, G.v.,   Gebieten zur Pruefung ihrer Eignung fuer die Warmwasser- Meske, C. and   Intensivhaltung. *Tierzuechter, 26*, H.5 Cellarius, O.,

Shehadeh, Z.H.,   1970:   *Controlled breeding of culturable species of fish — a review of progress and current problems*. FAO Indio-Pacific Fisheries Council, Bangkok.

Shehadeh, Z.H.,   1973:   *Induced breeding techniques — a review of progress and problems*. FAO/EIFAC Workshop on Controlled Reproduction of Cultivated Fishes, 21–25 May 1973, Hamburg.

Shelef, G.,   1978:   Photosynthetic biomass production from sewage. *Arch. Hydro-* Moraine, R. and   *biol. Beih.*, **11**, S.3–14. Oron, G.,

Shelton, L.,   1978:   *Use of Hormones to Produce Monosex Tilapia for Aquaculture.* Hopkins, K.D. and   Symposium on the culture of exotic fishes, Fish Culture Section, Jensen, G.L.,   American Fisheries Society, 257 pp.

Shigueno, K.,   1979:   Personal communication.

Siddall, S.E.,   1974:   Studies of closed marine culture systems *The Progressive Fish-Culturist, 36*, 1.

Sinha, V.P.R.,   1971:   Induced spawning in carp with fractionated fish pituitary extract. *J. Fish. Biol.*, **3**, 263–272 (1971).

Sklower, A.,   1951:   Carp breeding in Palestine. *Arch. Fischereiwiss.*, **3**, 42–54.

Soeder, C.J.,   1969:   Technische Produktion eiweißreicher Mikroalgen. *Umschau in Wissenschaft und Technik*, **69**, 801.

Soeder, C.J.,   1980:   Massive Cultivation of Microalgae: Results and Prospects. *Hydrobiologia, 72*, 197–209.

Soeder, C.J. and   1978:   *Microalgae for Food and Feed* — Ergebn. Limnol. Heft 11, E. Binsack, R.,   Schweizer-bartsche Verlagsbuchhandlung, Stuttgart, 300 pp. (Eds.)

Spannhof, L. and   1977:   Untersuchungen zur Verwertung verschiedener Futtermisch-Kuehne, H.,   ungen durch europaeische Aale (*Anguilla anguilla*) Arch. *Tierernaehrung*, 27, H.8, S.517–531.

Stanley, J.G.,   1976:   Female Homogamety in Grass Carp (*Ctenopharyngodon idella*) Determined by Gynogenesis. *J. Fish. Res. Board Can.*, **33**, 1372–1374.

Stanley, J.G. and   1976:   Feeding algae to fish. *Aquaculture, 7*, H.3, 219–223. Jones, J.B.,

Steffens, W.,   1957a: Die Wirkung von Hypophyseninjektionen auf Laichkarpfen. *Deutsche Fischerei-Zeitung, 4*, 83–87.

Steffens, W.,   1957b: Gewinnung und Injektion von Karpfenhypophysen. *Deutsche Fischerei-Zeitung, 4*, 265–272.

Steffens, W.,	1966:	Trockenfuttermittel als Alleinfutter fuer Karpfen. *Deutsche Fischerei-Zeitung*, **13**, 281–289.
Steffens, W. (Ed.),	1979:	*Industriemaessige Fischproduktion.* V E B Deutscher Landwirtschaftsverlag, Berlin.
Steffens, W. (Ed.),	1981:	*Moderne Fischwirtschaft — Grundlagen und Praxis.* Verlag Neumann-Neudamm, Melsungen.
Steffens, W. and Albrecht, M.-L.,	1976:	Untersuchungen ueber die Moeglichkeiten zur Senkung des Fischmehlanteiles im Futter fuer Regenbogenforellen (*Salmo gairdneri*). *Arch. Tierernaehrung*, **26**, H.4, 285–291.
Stoy, B.,	1981:	Fischfarm Limnotherm. *Bild der Wissenschaft*, **11**.
Stramke, D.,	1972:	Veraenderungen am Auge des europaeischen Aales (*Anguilla anguilla*, L. waehrend der Gelb-und Blankaalphase. *Arch. Fischereiwiss.*, **23**, H.2, 101–117.
Sugimoto, Y., Takeuchi, Y., Yamauchi, K. and Takahashi, H.,	1976:	Induced Maturation of Female Japanese Eels (*Anguilla japonica*) by Administration of Salmon Pituitaries, with Notes on Changes of Oil Droplets in Eggs of Matured Eels. Reprinted from *Bulletin of the Faculty of Fisheries, Hokkaido University*, **27**, 3–4, 107–120.
Suworow, E.E.,	1948:	*Grundlagen der Ichthyologie* (in Russian). Moskau, Woynarovich and Kausch, 1967.
Sylvester, J.R.,	1975:	Biological considerations on the use of thermal effluents for finfish aquaculture. *Aquaculture*, **6**, H.1, 1–10.
Takahashi, H. and Sugimoto, Y.,	1978:	A spontaneous hermaphrodite of the Japanese eel (*Anguilla japonica*) and its artificial maturation. *Jap. J. Ichthyol.*, **24**, 239.
Takeuchi, T., Yokoyama, M., Wanatabe, T. and Ogino, C.,	1978:	Optimum Ratio of Dietary Energy to Protein for Rainbow Trout. *Bull. Jap. Soc. of Scientific Fisheries*, **44** (7), 729–732.
Tanaka, J.,	1976:	*Utilization of heated discharge water from electric power plants in aquaculture.* FAO Technical Conference on Aquaculture, Kyoto. FIR: AQ/Conf/76/E.18.
Tesch, F.-W.,	1973:	*Der Aal.* Paul Parey.
Thielen, R.v. and Grave, H.,	1976:	Der Einsatz von Miesmuscheln (*Mytilus edulis* L.) in der Aquakultur. *Arbeiten des Deutschen Fischerei-Verbandes*, H.19, 33–41.
Thomas, W.H.,	1966:	On denitrification in the northeastern tropical Pacific Ocean. *Deep-Sea Research*, **13**, 1109–1114.
Tiews, K. (Ed.),	1981:	*Aquaculture in heated effluents and recirculation systems.* 2 vols. Heenemann, Berlin.
Tiews, K., Gropp, J. and Koops, H.,	1976:	On the development of optimal Rainbow Trout pellet feeds. *Arch. FischWiss.*, **27**, Beih. 1, 1–29.
Tiews, K., Koops, H., Gropp, J. and Beck, H.,	1978:	*Compilation of Fish Meal-Free Diets Obtained in Rainbow Trout* (Salmo gairdneri). *Feeding Experiments at Hamburg (1970–1977/78).* Proc. World Symp. on Finfish Nutrition and Feed Technology, Hamburg.
Tiews, K., Koops, H., Beck, H., Schwalb-Buhling, A. and Gropp, J.,	1981:	*Entwicklung von Ersatzfuttern fuer die Regenbogenforelle.* Veroeff. Inst. Kuest.-u. Binnenfisch., Hamburg. 77 pp.
Tiews, K., Manthey, M. and Koops, H.,	1981:	The carry over of fluoride from krill meal pellets into rainbow trout (*Salmo gairdneri*). *Arch. Fisch. Wiss.*
Uchida, R.N. and King, J.E.,	1962:	Tank culture of *Tilapia*. *Fishery Bull. 199*, **62**, 21–52.

Valenti, R.J.,	1975:	Induced polyploidy in *Tilapia aurea* (Steindachner) by means of temperature shock treatment. *J. Fish Biol., 7*, 519–528.
Vasilev, V.P., Makeeva, A.P. and Ryabov, I.N.,	1975:	On the triploidy of remote hybrids of carp (*Cyprinus carpio* L.) with other representatives of Cyprinidae. *Genetica,* **11**, H.8, 49–56.
Viehl, K.,	1940:	Der Einfluß der Temperatur auf die Selbstreinigung des Wassers unter besonderer Beruecksichtigung der bakteriologischen Verhaeltnisse. *Zeitschrift fuer Hygiene, 122*, 81–102.
Viehl, K.,	1949:	Die Umsetzung des Stickstoffs bei der biologischen Abwasserreinigung und bei der Schlammfaulung. *Gesundheits-Ingenieur*, Heft 21/22 (70. Jahrg.).
Vinogradov, V.,	1976:	Neue Objekte der Fischzucht und die Waermeenergetik (in Russian). *Rybovodsto i Rybolovstvo, Moskva,* **6**, 3–4. See: Merla, G. *Zeitschr. Binnenfischerei DDR,* **23**; H.4, 125–128.
Viola, S.,	1975:	Experiments on nutrition of carp growing in cages, part 2: partial substitution of fish meal. *Bamidgeh,* **27**, H.2, 40–48.
Viola, S., Mokady, S., Rappaport, U. and Arieli, Y.	1981:	Partial and complete replacement of fishmeal by soybean meal in feeds for intensive culture of carp. *Aquaculture,* **26**, 223–236.
Vogt, F.,	1978:	*Use of Solar Energy in Fish Farming.* Report no. 3, Polytechnic of Central London.
Vogt, F.,	1979:	*The application of existing solar energy collecting equipment in agriculture, horticulture and fishculture.* Proc. UNU Int. Sem. on Solar Technology in Rural Settings, Atlanta, Ga., USA. UNU, Tokyo.
Vogt, F.,	1979:	*The application of solar energy in agriculture, horticulture and fishculture.* Proc. Int. Conf. ISES.
Vogt, F.,	1980:	*The Use of Solar Energy in Agriculture, Horticulture and fishculture.* Report No. 10, Polytechnic of Central London.
Vogt, F.,	1981:	*Storage and use of solar energy in the bio-industries.* Proc. Int. Sem. Energy Conservation and Use of Renewable Energies in the Bio-industries, London. Pergamon Press, Oxford.
Vogt, F.,	1982:	*Fish Farming: its possible future structure.* Proc. 2nd Int. Sem. Energy Conservation and Use of Renewable Energies in the Bio-industries, Pergamon Press, Oxford.
Walter, E.,	1931:	Das Wachstum zurueckgehaltener Karpfen. *Fischereizeitg.,* **34**.
Watson, B.,	1965:	Characteristics of a Marine Nitrifying Bacterium, *Nitrosocystis oceanus* sp.n. *Limnol. and Oceanog.,* **10**, R274–7R289.
Weatherley, A.H.,	1976:	Factors affecting maximization of Fish growth, *J. Fish. Res. Board Can.,* **33**, 1046–1058.
Wedemeyer, G.A., Nelson, N.C. and Smith, C.A.,	1978:	Survival of the Salmonid Viruses Infectious Hematopoietic Necrosis (IHNV) and Infectious Pancreatic Necrosis (IPNV) in Ozonated, Chlorinated, and Untreated Waters. *Journal of the Fisheries Research Board of Canada,* **35**, 6.
Westernhagen, H. von and Rosenthal, H.,	1975:	Rearing and spawning siganids (Pisces: Teleostei) in a closed seawater system. *Helgolaender wiss. Meeresunters.,* **27**, 1–18.
Wild Jr., H.E., Sawyer, C.N. and McMahon, T.C.,	1971:	Factors Affecting Nitrification Kinetics. *Journal WPCF,* **43**, 9.
Wilkens, N.P.,	1981:	The Rationale and Relevance of Genetics in Aquaculture: An Overview, *Aquaculture,* **22**, 209–228.
Willer, A.,	1928:	Der Raumfaktorkomplex in der Forellenzucht. Mitteilung des Fischereivereins f. d. Prov. Brandenburg, *Ostpreußen und die Grenzmark,* **20**, 131.
Wohlfarth, G.W.,	1977:	Shoot carp. *Bamidgeh,* **29**, 2, 35–56.
Woynarovich, E.,	1953:	Die kuenstliche Fortpflanzung des Karpfens. *Act. Agr. Acad. Sci. Hung.* 3, 4.

'Woynarovich, E.,   1961:   Ausreifen von Karpfenlaich in Zugerglaesern und Aufzucht der Jungfische bis zum Alter von 10 Tagen. *Allg. Fischerie-Ztg.*, **86**, 22.

Woynarovich, E.,   1964:   *Ueber die kuenstliche Vermehrung des Karpfens und Erbruetung des Laiches in Zuger-Glaesern.* Wasser und Abwasser, Beitraege zur Gewaesserforschung IV, 210–217.

Woynarovich, E. and   1967:   Hypophysierung und Laicherbruetung bei Karpfen. In: *Vortragsveranstaltung ueber neue Methoden der Fischzuechtung und -haltung am 15. Februar 1967.* Selbstverlag Max-Planck-Institut Kausch, H.,   fuer Kulturpflanzenzuechtung, Hamburg.

Woynarovich, E. and   1980:   Modified technology for elimination of stickiness of common Woynarovich, A.   carp (*Cyprinus carpio*) eggs. *Aquacultura Hungarica*, **2**, 19–21.

Wuhrmann, K.,   1957:   Die dritte Reinigungsstufe: Wege und bisherige Erfolge in der Eliminierung eutrophierender Stoffe. *Schweiz. Z. Hydrol.*, **19**, 409–427.

Wunder, W.,   1956:   Leistungspruefungsversuche beim Karpfen 1955. Ergebnisse aus bayerischen Teichwirtschaften. *Arch. Fischereiwiss.*, **7**, 17–30.

Wunder, W.,   1960:   Erbliche Flossenfehler beim Karpfen und ihr Einfluß auf die Wachstumsleistung. *Arch. Fischereiwiss.*, **11**, 106–119.

Wunder, W.,   1966:   Beobachtungen und Betrachtungen ueber das Laichen der Karpfen. *Allg. Fischerei-Zeitung*, **91**, 17.

Yamamoto, K.,   1974:   Histological change of the hypothalamo-hypophysial system of Kasuga, S. and   the Japanese eel during maturation induced by Synahorin Omori, M.,   injection. *Bulletin of the Japanese Society of Scientific Fisheries*, **40** (2), 159–165.

Yamamoto, K.,   1974:   Artificial maturation of female Japanese eels by the injection of Morioka, T.,   salmonid pituitary. *Bulletin of the Japanese Society of Scientific* Hiroi, O. and   *Fisheries*, **40** (1), 1–7. Ommori, M.,

Yamamoto, K.,   1974:   Oogenesis of the Japanese eel. *Bulletin of the Japanese Society* Omori, K. and   *of Scientific Fisheries*, **40** (I), 9–15. Yamauchi, K.,

Yamamoto, K. and   1974:   Sexual maturation of Japanese eel and production of eel larvae Yamauchi, K.,   in the aquarium. *Nature*, **251**, 5472, 220–222.

Yamamoto, K.,   1975:   Pre-leptocephalic larvae of the Japanese eel. *Bulletin of the* Yamauchi, K. and   *Japanese Society of Scientific Fisheries*, **41**, H.1, 29–34. Morioka, T.,

Yamauchi, K.,   1976:   Cultivation of larvae of Japanese eel. Reprinted from: *Nature*, Nakamura, M.,   **263**, 5576, 412. Takahashi, H. and Takano, K.,

Yanzito, R.A.,   1981:   *The application of solar hot water and geothermal principles to closed cycle aquaculture.* Proc. Int. Sem. Energy Conservation and the use of renewable energies in the bio-industries. Pergamon Press, Oxford.

Zobel, H.,   1976:   Warmwasseranlagen zur Fischproduktion in der UdSSR. *Zeitschrift Binnenfischerei DDR*, **23**, H.3, 92–94.

Zweig, R.D.,   1981:   *Solar Aquaculture: An Ecological Approach to Human Food* Wolfe, J.R.,   *Production.* Bio-Engineering Symposium for Fish Culture, Todd, J.H.,   (FCS Publ. 1): 210–226. Engstrom, D.G. and Doolittle, A.M.,

INDEX